Canals and Rivers
of Britain

Canals and Rivers
of Britain

ANDREW
DARWIN

J. M. DENT & SONS LTD
LONDON

First published in 1976
© Text, Andrew Darwin, 1976

Printed in Great Britain by
Morrison & Gibb Ltd, London and Edinburgh
and Bound at the
Aldine Press, Letchworth, Herts
for
J. M. DENT & SONS LTD
Aldine House, Albemarle Street, London

This book is set in 12 on 13pt Poliphilus

ISBN: 0 460 04207 6

Contents

Acknowledgements

My thanks to Richard Murby, David Perrott and Derek Pratt for contributing photographs to supplement my own; to David Perrott for putting together the maps; and to my wife, Marie-Hélène, without whose assistance and encouragement this book would not have seen the light of day. Maps reproduced by kind permission of John Bartholomew & Son Ltd, Edinburgh. Derek Pratt's photographs are on pages 34, 35, 38, 39, 42, 52, 53, 54, 55, 65, 66, 67, 75, 76, 82, 89, 90, 112, 120, 138, 139, 190, 204.

Author's Foreword

This book is for anyone interested in the inland waterways of England, Scotland and Wales. The ever-growing numbers of modern-day explorers who navigate our canals and inland rivers for pleasure should find it useful for planning their next cruise, while those who want to do a little weekend canalling, or to seek out a good canal walk or an interesting flight of locks should also be helped on their way. And anyone else who cares to discover more about the once-hidden world of the waterways should find plenty of fascinating material here.

The aim is to provide a readable guide rather than a mere gazetteer, and enough history to help towards an understanding of the context and importance of each waterway. The maps are intended to make the book as practical as possible and to show the position, extent and connections of each waterway. The pictures explain themselves.

The accent is very much on *navigable* waterways, and canals which have long been abandoned or even filled in are not accorded much space in these pages, although they may be mentioned in passing. But consistency has been hard to maintain. The *Kennet and Avon Canal*, for example, is still officially navigable only in parts, and yet it is 'in water' for virtually the whole of its length—as well as being probably the longest and best-known artificial waterway in southern England. Clearly, it would have been unrealistic and undesirable to leave it out. More difficult candidates were such as the *Stour Navigation* in Suffolk, which was ignored because it is unnavigable, short and of little significance. Its neighbour the *Chelmer and Blackwater Navigation,* like the *Stour* is connected only to the sea, but it is still fully navigable and in use and was thus judged deserving of an entry.

A final word on the use of the book. The arrangement of the canals and rivers is fundamentally geographical in that they are grouped by regions. Thus, the book can either be read from beginning to end, or referred to for particular sections describing certain stretches of waterway. In any case, the maps, table of contents and index all help to make the book easy to use—for quick reference or for the sheer pleasure of armchair reading and viewing.

I

The Midland Framework

TRENT AND MERSEY CANAL

CALDON CANAL

STAFFORDSHIRE AND WORCESTERSHIRE CANAL

OXFORD CANAL

COVENTRY CANAL

BIRMINGHAM AND FAZELEY CANAL

ASHBY CANAL

This group of canals forms the geographical and historical nucleus of the English canal network. These navigations include the framework of early canals connecting England's four principal river basins—Humber, Thames, Severn and Mersey. These enclose the heart of England and the bulk of its industrial wealth. Most of them were built in the first era of canal construction in this country, and all except the *Ashby* were designed to the narrow gauge which saved on construction and water supply costs but which, never subsequently modified, was eventually defenceless against the competition from first the railway and later the motor lorry.

Trent and Mersey Canal

It is hard today to imagine the one-time significance of the *Trent and Mersey Canal,* but this little water byway was once the 'Grand Trunk' of the waterway system, and its conception as a waterway had an immediate and profound impact upon entrepreneurs throughout Britain. It linked numerous inland towns and villages to both the Mersey and Trent river basins and ports, thereby enabling goods to be moved swiftly, reliably and cheaply from one

The junction of the rivers Trent and Derwent with the *Trent and Mersey,* near Long Eaton, in Derbyshire. In the middle distance, the Sawley Cut bypasses one of the Trent weirs.

place to another; and, in conjunction with the *Staffordshire and Worcester-shire Canal,* it laid the foundations of the whole complicated network of narrow canals in the Midlands. Although both canals found themselves sooner or later assailed by competition from not only the railways but other canals, each has still managed to survive intact, and indeed to continue as a useful transport waterway until quite recently, despite their obvious physical limitations.

The *Trent and Mersey Canal* was originally promoted by the potter Josiah Wedgwood, who had observed the construction of the Duke of Bridgewater's canal from Worsley to Manchester. Wedgwood realized the potential benefits that such a waterway could bring to the pottery industry by greatly facilitating the import of raw materials from far away and, at the same time, enabling finished goods to be despatched to their markets in reliable and steady vehicles with a minimum of breakages. He was not short of influential and energetic friends, and, by 1766, a Parliamentary Bill had been passed, and the engineer of the *Bridgewater Canal,* James Brindley, had been enlisted. Work started without delay.

The *Trent and Mersey* scheme was no mean undertaking. The line chosen by Brindley included seventy-six locks and five tunnels, including one of 2900 yards through Harecastle Hill. Derision greeted the news of this part

'Paired' narrow locks on the *Trent and Mersey Canal*.

of the project, just as Brindley's plan to build a canal aqueduct over the Mersey had been derided; but although Brindley never lived to see it, the great tunnel was duly opened in 1777. When, less than fifty years later, it was badly affected by subsidence caused by local mining, Thomas Telford was engaged to resolve the problem. He chose to build a second one alongside the first, and the two tunnels are still there, although Brindley's is now closed altogether and Telford's has also begun to show its age and requires heavy expenditure on maintenance.

The canal starts at the junction of the rivers Trent and Derwent, at a point about seven and a half miles south-east of Derby. (The fact that Brindley chose to ignore the river route already existing as far upstream as Burton-on-Trent was the cause of much regret to the owners of the *Upper Trent Navigation.*) The canal locks steadily up the valley: wide locks to Burton to allow river craft access thus far, and narrow locks thereafter. At Fradley is a junction with the *Coventry Canal,* and here the *Trent and Mersey* turns north-west for Stoke, being joined by the *Staffordshire and Worcestershire* near the village of Great Haywood. At Stoke, the canal climbs steeply to its summit in Etruria before penetrating the ridge of

An outstanding relic of the canal age at Shardlow, on the *Trent and Mersey Canal*.

Harecastle Hill. It emerges at Kidsgrove and immediately begins the descent through Cheshire. At Middlewich (a well established salt mining district), it picks its way up the valley of the River Dane and follows this to the River Weaver: it runs beside and above the Weaver, passing through two tunnels north of Northwich before striking northwards to go through the tunnel at Preston Brook and make an end-on junction with the *Bridge-water Canal* at a point overlooking the Mersey estuary.

The *Trent and Mersey* is a canal of contrasts, passing as it does from the soft water meadows—and dominant power stations—of the Trent valley to the very heart of the Potteries, where old kilns, canal basins and abandoned arms are all to be found in abundance. The canal even flows through the middle of Shelton Steelworks. There is the drama of Harecastle Tunnel, followed by its passage through the rich farmlands of Cheshire before its encounter with Britain's major salt mining district around Middlewich, with its legacy of drastic land subsidence. And finally there is the journey beside the Weaver valley, the canal diving in and out of narrow tunnels before emerging on to the side of the Mersey valley near Runcorn. There is contrast in the locks too: the wide ones at either end (downstream of Burton and Middlewich) and the narrow ones in the middle, with pairs being used throughout most of Cheshire and even a unique steel-framed one (Thurlwool Steel Lock). There is also the unique Anderton Boat Lift, connecting the canal with the Weaver fifty feet below.

PLACES TO SEE THE CANAL

It may be evident from what has been said that there are endless points of interest along this canal, the highlights being the Anderton Lift and Harecastle Tunnel. Other good places to capture its flavour include Stone, a small country town served by the canal with a flight of locks, old and new canal boatyards, and a good canal pub called 'The Star'.

A walk from the centre of Stoke-on-Trent and north for a couple of miles will show what the canal was really built for. Further south, there is an early Brindley aqueduct to carry the canal across the Trent at Rugeley;

More 'paired' locks on the *Trent and Mersey*. This arrangement was used to save both time and water in busier days. Today, the cost of maintaining both locks is prohibitive.

Harecastle Tunnels. The older one (on the right) is now closed, while Telford's newer one is also showing signs of age, and headroom is restricted.

Anderton Lift. It was built to last.

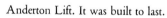

and just west of nearby Armitage there used to be a short tunnel which was only opened out in 1971. Fradley and Great Haywood Junctions are both of great interest, the former for locks, boating activity and the range of canal buildings focused on the 'Swan Inn'; and Great Haywood for the delicate poise of the junction roving bridge and its position near the junction of the rivers Trent and Sow with Shugborough Hall just over the water. The *Staffordshire and Worcestershire's* aqueduct over the Trent—and the unlikely Tixall Broad—are only a stone's throw away.

But it is the Anderton Lift and Harecastle Tunnel which are inevitably of the greatest interest. The tunnel can only be properly appreciated by boat, of course, although visitors to either end will see the now almost-submerged entrance to Brindley's tunnel; and they can see from the loading gauge that hangs above the entrance to the 'new' one (it was completed in 1827) the problems that both tunnels have suffered. The approach to the north end of the tunnel is in any case a good area for canal exploring. The junction with the *Macclesfield Canal* is nearby—an unusual arrangement involving a 'flyover', the canals running parallel for a while before one crosses the other on an aqueduct to avoid locks.

The Anderton Lift is unique in Britain, and the biggest piece of machinery on the canals. It was built to connect the *Trent and Mersey Canal* to the *Weaver Navigation* at a point where the canal is right beside but fifty feet above the river. It is a huge, complicated affair of rows of iron girders, struts and columns. There are two caissons or tanks, once connected hydraulically but now independent and electrically operated. The boats simply sail in, watertight gates are closed and the tank moves vertically down to let the boat out into the river. It is a marvellous device and is easily found, overlooking the Weaver just west of the village of Anderton.

BOATING ON THE TRENT AND MERSEY CANAL

Taking a boat on the *Trent and Mersey* offers the chance to see some of the extremes in canal environments—from the peaceful solitude of the Dane valley to the very heart of the Potteries, with a trip through a steelworks thrown in. There are plenty of locks, especially through Cheshire, and few pitfalls. Harecastle Tunnel, however, has suffered from subsidence, and headroom is limited to that shown by the loading gauge at each end. Since the tunnel is much too long to see through, boats may only enter during given periods. A tunnel keeper controls access.

Dredging and new steel piling were two of the many kinds of work involved in the *Caldon*'s recent restoration to navigability.

Caldon Canal

The Caldon was built as a branch of the *Trent and Mersey;* it has now rightly acquired an identity of its own. It is one of the most beautiful and secluded of all the canals, and although it has been virtually abandoned and largely un-navigable for some years, local enthusiasm and effort combined with forward-looking official opinion have ensured that the canal has now been restored and is fully navigable once more.

The canal runs from Etruria, Stoke-on-Trent, to the nearby Churnet valley which it follows down to an isolated industrial settlement at Froghall. It was constructed by the *Trent and Mersey* company mainly to allow access to the big deposits of limestone at Caldon Low, in the hills near Froghall.

The branch was opened as early as 1779, a line seventeen and a half miles long. By 1802, a subsidiary branch had been built to a new reservoir near Leek (Rudyard Lake), and not long afterwards a thirteen mile extension was opened from Froghall to Uttoxeter. The latter did not last long—less than forty years—but the main line of the *Caldon* was well supported by the tramways which brought the limestone down from the Caldon quarries to Froghall Basin, and the traffic proved the canal's mainstay throughout its working life. The Uttoxeter Branch has disappeared almost entirely; the Leek Branch now stops just short of the town, and the canal layout between Endon and Denford underwent several changes up to 1841; but otherwise the *Caldon* remains as built.

Most people who discover the canal for the first time are amazed by its inaccessibility—a drawback that very soon becomes one of its greatest assets to those already on it. But the canal is anyway full of surprises—and unexpected pleasures—for the canal explorer. There are lift bridges, a short tunnel, a canal flyover, tiny remote pubs and even a pair of restored, flint watermills right beside the waterway. And most of it is framed by one of Staffordshire's most hidden valleys. It is a marvellous little canal, so that one can hardly imagine it ever being allowed to be abandoned.

The canal starts at Etruria Top Lock in the heart of the old area of the Potteries. Josiah Wedgwood's original pottery stood not far to the north of

Bridge detail at Endon, on the *Caldon Canal.*

Etruria Top Lock, in the heart of the Potteries, with a maintenance dock to the left. Until recently, both were fully roofed.

the junction. The canal immediately zig-zags away up two staircase locks before circling discreetly through Hanley and out the other side. It winds through Foxley where there are, somewhat unexpectedly, two little bascule lift bridges. Nearby the canal's only two carrying craft are based—modern pontoons built specially for local transport use on the *Caldon Canal* by a waterside pottery company. The canal crosses the headwaters of the River Trent. Six more locks take it up to summit level, over a saddle and down to the Churnet valley. At Denford there is a fork, with locks going down ahead and a branch off to the right. This is the start of the Leek Branch, which was built primarily to act as a feeder for the canal from Rudyard Lake to *Caldon's* summit level. The last mile into Leek has been infilled now, but the feeder is naturally still very much open (the supply is also an important source for the *Trent and Mersey's* summit level). The other section of the waterway—the part which was re-opened only in 1974—locks down to the Churnet, a river running south-east through a beautiful and steeply-wooded valley where there are no roads, just a single line of lightly-used railway track. It is this valley which is the true glory of the *Caldon Canal*. But although sparsely inhabited, it is by no means empty. At the village of Cheddleton there are two magnificent water mills built to grind flint for the pottery industry. They have both been restored. (Further down there is a more modern flint-mill—also water-powered—which is still used.) Mean-

The *Caldon Canal* in the remote Churnet Valley. The design of the flimsy iron and timber footbridge at the lock indicates the canal's *Trent and Mersey* parentage—and contrasts strongly with the substantial nature of the bridge in the foreground.

while, the navigation joins the actual river course for a mile to Consall Forge, a tiny riverside settlement tucked away in this hidden valley. There is also, unbelievably, a pub here, although most of its trade has probably more to do with serving tea to local climbers and walkers than with catering for evening drinkers. From here the canal and river separate again, the former winding away along the side of the steeply-wooded hillside before reaching the lonely but apparently thriving industrial settlement of Froghall. There is a short, low tunnel here before the canal ends at a small, Y-shaped basin.

PLACES TO SEE THE CANAL

The west end of the *Caldon Canal* in Etruria is, of course, intensely industrial, although changes in the pottery industry have left Etruria a distinct backwater. The age and condition of the nearby factories and works reflect this change. But much more typical of the *Caldon* is the other end, to be seen in the wooded Churnet valley. Principal areas of interest on the canal are Stockton Brook, where the five-lock rise renders fine views over the heavily built-up valley of the Trent headwaters; Denford, with some splendid locks and an intriguing canal aqueduct flyover; and Cheddleton, with more locks and a pair of old water wheels at the flint-grinding works. Froghall is also of interest, with its sheltered basin and old lime kilns, while the remains of the Uttoxeter Branch's first lock are clearly visible nearby. But in between these places is a canal of great beauty. It is well worth a good, long walk.

21

Staffordshire and Worcestershire Canal

This canal is a direct contemporary of the *Trent and Mersey* and was also engineered by Brindley, although its opening in 1772 was five years sooner. Its provision of an early West Midlands–River Severn link, and its junction with the *Trent and Mersey* at Great Haywood, established it as a primary route in the structure of North-West–South-West trade, and its existence as the shortest line available outweighed its somewhat exaggeratedly twisting course.

Like the *Trent and Mersey,* the *Staffordshire and Worcestershire* (or 'Staffs and Worcs' as it is known) is a very small, narrow winding waterway of considerable interest and (in places) of exceptional beauty. Much of the interest lies in the early engineering aspects of the canal. Brindley built his first lock on this canal (Compton Lock), and the well-known circular weirs at many of the locks represent a piece of experimental design in weir construction. There is an early Brindley aqueduct near Great Haywood. Other noteworthy touches are the octagonal toll offices and the split bridges at the tail of some of the locks, and a cantilevered iron bridge in two sections, divided by a mere inch or two to allow a towing line to be dropped through without disconnecting the horse from its tow. There are some curious locks too, like the arrangement at the Bratch; and there are some delightful names— Tixall, Stewponey, Bumblehole and Dimmingsdale locks, as well as Giggetty, Long Moll's and Weeping Cross bridges. The latter names are often boldly set out on ancient cast-iron name plates, although these are now sadly diminished in number. But it is not just the canal itself and its paraphernalia that are so appealing, for its route through its surroundings is truly enchanting. Except for a short stretch in the middle, the canal seems to stick to small river valleys which are sometimes open and sweeping, but more often narrow, wooded and intimate. Even where the waterway is less than two miles from the centre of Wolverhampton, the canal manages to take refuge in a cutting that successfully shields it from the modern world a few yards away on either side.

The canal leaves the *Trent and Mersey* at Great Haywood Junction, crossing the River Trent on an aqueduct and joining the Sow valley, a broad valley forming the north side of Cannock Chase. The canal widens out at Tixall Broad—an unexpected 'bulge' in the waterway—then begins to lock up the valley, crossing the River Sow on a low, heavy aqueduct before heading southwards near Stafford to exchange the Sow for the River Penk. More locks lead the canal up past Penkridge to the ten mile summit level that starts at Gailey. North of Wolverhampton is a shallow cutting only

Fresh paint at Bratch Locks on the *Staffordshire and Worcestershire*.

A secluded lock on the *Staffordshire and Worcestershire Canal*, at Whittington.

Stewponey Lock, near Kinver, on the *Staffordshire and Worcestershire Canal.*

wide enough for a single boat; and then there are important canal junctions within a half a mile of each other linking the *Staffs and Worcs* to the *Shropshire Union* and *Birmingham* canals.

At Compton is the first of thirty-one locks down towards the River Severn at Stourport. The country on either side begins to get a little hillier as the canal approaches a great sandstone ridge, and from here on it follows a tortuous and secluded course, mostly through the thickly wooded valley of the River Stour, which is hemmed in by steep sides of sandstone rock. Picturesque locks and canalside pubs alternate with great red sandstone cliffs overhung with foliage rising sheer from the waterway. There are tiny tunnels at Cookley and Dunsley, and a remarkable wall of rock leaning right out over the canal at Austcliff. At Kidderminster, it drops down a lock in the town centre and disappears through a short tunnel before regaining open country and following the Stour Valley into Stourport. In the town, extensive basins and various locks separate the canal from the River Severn.

PLACES TO SEE THE CANAL

There are many good things to see on this canal. The northern section is rural and attractive in an undramatic, typically Midland sort of way, although the Tixall area is worthy of note, mainly for the scenery. There are some good locks between Penkridge and Gailey top lock, with an old canalside pub in the middle. The main reservoirs feeding the canal are not far away but are split by the M6 motorway, which has robbed several miles of the canal of its former peace.

The southern section is of great scenic and historical interest. Any part of the canal between Wombourn and Stourport is worth investigating, especially the area nearest to Kinver where the sandstone cliffs are at their most spectacular. South of Wombourn, Botterham Locks make a two-step staircase, and Bratch Locks to the north of the town are an intriguing threesome, in semi-staircase formation. Cookley is also a good base from which to see not only the little tunnel, but also Debdale Lock a short distance downstream. Out of the steep, red rocks that shelter the lockhouse a canal stable has been carved, and to the north of Cookley village is the extraordinary sight of Austcliff Rock.

Easily the most famous place on the canal is Stourport, a town which was created purely by the arrival of the *Staffordshire and Worcestershire Canal* at the east bank of the river Severn. It is justly famous for its vintage canal architecture and the authentic flavour of the heyday of canal and river transport. There are basins, broad and narrow locks, drydocks, workshops and, of course, boats everywhere. The trading boats may have vanished, but pleasure craft do not ignore the facilities here, and Stourport continues to thrive.

25

The delights of this splendid waterway are well known, and in spite of it being part of an important route, the canal is not too heavily trafficked—which is just as well, for part of the canal's course is a long succession of blind hairpin bends, and it is hard to imagine it being well-used by narrow-boats without the occurrence of endless minor collisions. There are forty-three locks in the forty-six miles of waterway, most of which are fairly regularly spaced between Wolverhampton and Kidderminster. There are plenty of boating centres along the canal, including hire cruiser bases, and a trip boat operates from Kinver.

Oxford Canal

Because the *Grand Union Canal* shares the course of the *Oxford Canal* for five miles between Braunston and Napton, the *Oxford Canal* is clearly divided for most purposes into its northern and southern sections—a distinc-tion that gains usefulness from the pronounced difference in aspect of the two sections. The northern part, from Hawkesbury Junction to Braunston, is very much a part of the Midlands. Hawkesbury itself, on the edge of Coventry, has a huge power station breathing down its neck, and a few miles further south is Rugby, an engineering town which is skirted by the canal as it follows the hillside opposite. The land south of here is flat, open and a little empty.

Such industrial manifestations are a far cry from the atmosphere on the southern reaches of the *Oxford Canal,* for the latter traverses countryside that is nothing if not rural, peaceful and utterly unspoilt. It is a gentle land of soft, rolling fields and discreet ridges of green hills. The canal visits one village after another, and everywhere are the warm, brownstone cottages and farms that represent the best of Oxfordshire. Shy little brick bridges alternate with tiny wooden lift bridges, while a steady but rarely steep succession of narrow locks leads the canal on down the Cherwell valley towards the River Thames. Watermills or their sites can be found beside some of the locks, and for many lengths trees overshadow the canal. At one point, the canal actually locks down into the Cherwell for a mile or so before leaving it at a rare, diamond-shaped lock.

There are other conspicuous differences between the two sections of the

An iron lift bridge over the *Oxford Canal* at Lower Heyford.

canal. Up at the north end the villages are of midlands brick, not brown-stone, and the landscape wears a somehow resigned and world-weary air by contrast to the unsullied innocence of the south. There are more roads and more traffic, more buildings and more people. On the canal itself, the only locks are the flight of three pairs at Hillmorton, and the only rivers are crossed on aqueducts and quickly left behind. The little black and white wooden lift bridges that are such a well-known feature of the southern *Oxford Canal* are nowhere to be found on the northern part, whose own trade mark is rather the balustraded iron bridges that were erected as a result of the improvement scheme of the early 1830s.

The *Oxford Canal* was built fairly early in England's canal age. It was authorized in 1769, only three years after the *Trent and Mersey,* and these two canals, with the *Coventry* and *Staffordshire and Worcestershire,* were each to form a principal limb in the framework of waterways connecting the country's four main river basins. Their intersection in and around the West

'Braunston Turn'—the 'new' junction of the *Grand Union* and *Oxford* canals. The northern half of the *Oxford Canal* is as easily recognized by its cast iron bridges as is the southern half by its wooden lift bridges.

Midlands, and the construction of the *Birmingham Canal,* gave this region a flying start in the Industrial Revolution.

The *Oxford* was envisaged as a lone southward extension from the *Coventry Canal* (authorized a year before) near Coventry. It was aimed at not only bringing Midlands coal down to Banbury and Oxford but at opening up, via the River Thames, the first Midlands–London water link.

As built, the canal—whose general line was the responsibility of James Brindley—was a full ninety-one miles from the *Coventry Canal* to Oxford. Much of this length was unnecessary, partly because of unhelpful company rivalry which led to a mile of route duplication at the northern terminal of Longford Junction, but mostly because of the strict contour-hugging line that was followed on the northern two-thirds of the route. This could imply a very conservative approach to the engineering of the canal, or simply the desire to serve as many villages and farms as possible. Or perhaps it reflected the confidence of a company that was broaching fresh ground in a region as yet devoid of competition. But the *Oxford Canal* was not opened throughout until 1790, and very shortly the plans for a new route from Braunston to Brentford (the *Grand Junction Canal*) upset the *Oxford's* monopoly position. However, it was not until 1829 that the Oxford Canal Company finally got around to straightening their line north of Braunston. An energetic and comprehensive scheme was undertaken, shortening the route by almost fourteen miles. This was completed in 1834 at a cost of over £160,000.

The improvement scheme was responsible for drastic changes to the face of the canal. Shortening the route transformed it from a narrow, winding

In the last stages of decay, an 'accommodation' bridge at the *Oxford Canal*'s summit level. Many of the minor bridges that span canals were built simply to reconnect fields divided by the new 'cut'.

waterway of unambitious appearance to a much more business-like route with plenty of cuttings, embankments and aqueducts. The more isolated lengths of the old route, which wandered round the hillsides and valleys instead of cutting through or over them, have virtually vanished into the fields in most places, but the intersections with the new line have often become branches of varying usefulness, and it is here that the 'Horseley Ironworks 1828' bridges are to be found, giving views of anything from lines of moored pleasure boats to a weed-choked dead-end. Only the extraordinarily devious line of the summit level between Claydon and Marston Doles survives as a vivid reminder of what the northern section of the *Oxford Canal* used to be like.

PLACES TO SEE THE CANAL

Because of the canal's somewhat easy course through the countryside, the *Oxford Canal* does not contain many spectacular structures, although there are plenty of embankments and cuttings on the northern section. The attraction of the canal is rather its gentle, sleepy character and the quality of the surrounding countryside than anything spectacular.

There are no towns along the canal except for Rugby, which is across a valley from it, and Banbury, which has spurned the canal that runs close to the heart of the town. An old canal basin has been filled in and turned into an ugly bus station, while old mills and virtually all the other nearby buildings of any age or character have been demolished. No recognition is made even today of the contribution that the canal could make to the urban scene if it was allowed to; but the villages along the canal are delightful, particularly in the Cherwell valley, where it is horrifying to contemplate the suggested routing of the M40 motorway past the quiet and unspoilt brownstone villages of the upper Cherwell valley. Such a road would destroy the magnificent situation of villages like Somerton and King's Sutton, with its incomparable church spire. The character of the canal too would be completely changed for several miles.

PLACES TO SEE THE CANAL ON THE NORTHERN SECTION

HAWKESBURY JUNCTION: This scruffy little hamlet, in the very shadow of a large power station, is full of surprises, mainly because of modern failure to get to grips with the place. Hawkesbury is one of the few remaining canal communities which remains intact, and amongst the narrowboats tied up at this unlikely spot is a nucleus of boat-bound residents, old canal boatmen—and boatwomen—who have retired with their boats to their particular idea of home. The junction itself is between the *Oxford* and *Coventry* canals, and has a still-operational stop lock. This lock was doubtless insisted upon by the Coventry Canal Company to ensure that if any water flowed from one canal to the other, it could only be to the advantage of the *Coventry*. The iron 'roving' bridge over the junction is dated 1836. Prior to that time, the two canals ran parallel before joining together three quarters of a mile to the south-west. Also at Hawkesbury is an old engine house which used to pump water up into the *Coventry Canal;* and best of all is the 'Greyhound', a proper old canal pub.

NEWBOLD-ON-AVON: One of the many places where the northern *Oxford Canal's* line was straightened in the 1820s. There is no sign now of the junction, but it is worth noting the 'new' tunnel with its generous dimensions, twin towpaths, and a roving bridge incorporated into each end. Across the road past the two canal pubs near the sheltered wharf is Newbold Church. In the field beside the church is the entrance to the old tunnel—it is interesting to try and work out where the far end of this tunnel would have been.

HILLMORTON: Cut off by a railway embankment from Hillmorton itself is an interesting and busy canal settlement. There is a flight of three narrow

Thrupp—a typical *Oxford Canal* lift bridge.

locks, all paired, between which lies a group of old buildings. A canal maintenance yard explains the private-looking arm and drydock that lead off from the canal just above the lower locks. Note the old canal manager's house next door, and the tiny bridge leading to it.

NAPTON ON THE HILL: Napton village has established itself about the Hill, a great lump of a thing with a conspicuous windmill on top. The canal has to go right around this obstacle to climb the nine locks to its summit level before starting to fall towards the Thames valley, still another forty-five miles away. North of Napton Hill is Napton Junction, where the *Grand Union* and the *Oxford Canal* part company; a new boatyard has been excavated out of the fields nearby. At the bottom of the locks there are usually a few craft about. This used to be a popular overnight stop among the narrowboat crews, and the house opposite was used as a canal pub, the 'Bull and Butcher'. The locks themselves are a gentle flight, liberally dotted with weatherbeaten brick bridges, and fitting snugly into the fields

31

on either side. At the top of the flight sits Marston Doles, a minute hamlet with a good canal warehouse.

An oddity along the locks here is the old canal arm, abandoned but still in water. It used to run a short way to an engine house built to pump water back up Napton Locks to supplement the three reservoirs maintaining the summit level. The engine house has disappeared now, almost without trace. It has only just been replaced in fact—by a £50,000 set of back-pumping motors along the flight.

CLAYDON, CROPREDY AND UPPER HEYFORD: These are three beautiful, brownstone villages near the canal but clearly established long before it. Claydon is just west of the five locks that lift it onto its meandering summit level, while Cropredy sports just one lock, hard by the church and one of the village's two excellent pubs. Upper Heyford has the only cast iron lift bridge on the canal, right next to a large and ancient mill over the Cherwell. Weeping willows overhang the canal to complete the scene, which is in no way tarnished by the presence of a large air base just over the hill.

THRUPP: A small, linear canal settlement near Kidlington, Thrupp supports a pub, a row of unassuming canal cottages, a lift bridge and a maintenance yard. There are always boats about at this delightful spot.

OXFORD: The southern terminus of the canal is full of mystery. It is virtually unknown by most of the townspeople, for the canal is heavily disguised among Thames backwaters at the north end of Oxford Station. It is linked to the river by a small lock—savants will immediately recognize the tell-tale iron roving bridge which crosses it. The canal continues as a weedy water channel south of this lock for a short distance before being cut dead by a road. Back at the lock, the canal runs north past Oxford's back gardens under the first of many lift bridges (the second is a small electrically operated one) and out towards Kidlington and the peaceful world of the Cherwell valley.

BOATING ON THE OXFORD CANAL

This waterway is very well suited to and equipped for pleasure boats. There are plenty of boatyards and hire-cruiser bases, particularly on the immensely attractive southern section which has been a prime hire-cruiser waterway for years. Many of today's canal enthusiasts first caught the canal 'bug' on the southern *Oxford*. The lift bridges are mostly easy to work, although not all of them are entirely counter-balanced and it is as well to hold the balance beam while a boat is passing underneath. The locks are somewhat slow but easy to work, and few of them are deep enough to cause any difficulties— except for the twelve feet drop of Somerton Deep Lock which often catches people out the first time around. Other points in favour of the *Oxford*— particularly the southern section—are the number of pretty villages to visit nearby, and plenty of good pubs and mooring sites. The disadvantage is that the canal is often somewhat crowded with boats; it also tends to be rather shallow, especially towards the southern end.

Coventry Canal

The *Coventry Canal* looks very unpromising on a map. Its route through or past an apparently endless series of industrial and mining towns seems to be interrupted only by coal mines, quarries, old railways and power cables: one might be forgiven for wondering what a Coventry Canal Society would have to enthuse about. But a journey along the waterway shows this impression to be false, for the very presence of the mining industry and quarrying concerns explains not only the strange 'lunar' landscape formed by the old tips, but also the industrial relics and canal curiosities, and the rusting coal chutes and loading basins, all of which are quite intriguing. Nor is it difficult to see that the waterway was used commercially until relatively recently. But there are other aspects—the canal workshops at Hartshill, the supposedly haunted ruins of Alvecote Priory, the descent through Atherstone locks and a good measure of robust open countryside— all of which give the *Coventry Canal* its fair share of intrinsically attractive features.

The canal was partly conceived as a feeder of coal from the Warwickshire coalfield to the north into the bustling town of Coventry to the south, and partly to give Coventry and the region an outlet for 'exports' via the new *Trent and Mersey Canal* which it was to join at Fradley, near Burton-on-Trent. But the *Coventry* soon became an important part of various plans that would, between them, weave a network of canals around the industrial Midlands; and when the Coventry Canal Company was unable to finish construction because of shortage of cash, its neighbours undertook to build the last section from Fazeley to Fradley for everyone's mutual benefit. The Coventry Company later bought back the part of the line originally built by the Trent and Mersey Company (from Whittington to Fradley Junction), but never

33

Part of the maintenance yard at Hartshill.

got round to buying back the other section (from Fazeley to Whittington): theoretically, this remains part of the *Birmingham and Fazeley Canal,* leaving the more northerly length a detached part of the *Coventry Canal.*

Considering the route of the canal in more detail, it starts in a Y-shaped basin actually in Coventry, near the centre of the city, and twists between rows of factories before leaving past back gardens, a gas works, the M6 motorway and Hawkesbury Junction. Somewhat despoiled countryside leads up to and past the mining town of Bedworth—the *Ashby Canal* leads off to the east of this point. Various disused branches lead off to the west in

The British Waterways Board house by Fazeley Junction, the meeting place of the *Birmingham and Fazeley* and *Coventry* canals with the River Tame.

this area, and are fading reminders of the once extensive system of private colliery canals that pre-dated the *Coventry Canal* itself.

The canal dodges round Nuneaton to emerge onto the side of the Anker valley where an intensive quarrying industry, centred on the Hartshill area, gives the canal water a bright, rusty-red colour. There is a certain bizarre grandeur in the hilly landscape, with its great conical spoil heaps and long-dead light railways which cross the canal on flimsy iron bridges. They contrast strongly with the very much alive electric railway that races along the valley below. At Atherstone is a flight of eleven locks, steep at first, but becoming somewhat long drawn out. There are some delightful spots towards the lower end of the flight at the coincidence of the locks and the old brick bridges.

North of Atherstone the waterway follows the Anker valley round Polesworth and its venerable abbey, and past the mournful shell of Alvecote Priory. (The latter, a few yards from the canal, was the victim of the subsidence that has affected so much of this area and has caused extensive flooding in the valley.) The other two locks on the canal are encountered at Glascote, on the outskirts of Tamworth. An aqueduct over the River Tame leads to Fazeley Junction, and the rural and sometimes densely wooded stretch of canal that follows is the *Birmingham and Fazeley*. The *Coventry Canal* restarts at an unmarked spot near the village of Whittington and runs through flat countryside to Fradley Junction, passing at Huddlesford the junction with the old *Wyrley and Essington Canal*. This former, heavily-locked link with the canals of the Black Country is now completely defunct beyond the first bridge, a quarter mile down from Huddlesford Junction.

PLACES TO SEE THE CANAL

The *Coventry Canal* is by no means a spectacular waterway: rather, it is a very ordinary Midlands canal which rarely attracts more than scant attention. But there are some good brick bridges, plenty of old junctions and a variety of abandoned canal branches to explore in the Bedworth area. The towpath along the canal is generally in good condition.

FAZELEY: The A5 passes obliviously over the junction here between the *Coventry* and *Birmingham and Fazeley* canals. The old toll house is still occupied, and is numbered like all the other old company houses on the *Birmingham and Fazeley*. Half a mile to the east is a low aqueduct over the River Tame.

ATHERSTONE: A flight of eleven locks drops inconspicuously around the back of the town past the station. The old A5 bridge half way down the flight is a remarkably sheltered place—and popular among boaters for an overnight stop. Even further down is Baddesley Basin, which supplied coal for some of the last working narrowboats to use the canal in the 1960s.

HARTSHILL MAINTENANCE YARD: Neatly bracketed by two identical canal bridges is this fine example of early canal architecture which is still

36

used for its original purpose. A great, yawning arch (see Plate 00) gives access for boats into the unlit interior of the principal building, but the air of mystery is more than outweighed by the jaunty wooden clock tower on top.

GRIFF COLLIERY ARM: An old, three-quarter-mile-long private arm, this used to lead off towards a colliery just south of Nuneaton. Not just one but two roving bridges used to transfer the towing path from the *Coventry Canal* to the 'right' side of the branch. To the west is Arbury Hall, once the seat of the Newdigate family, the coal owners who constructed a small private network of canals to serve their collieries in this area. It is a good area for canal explorers.

BOATING ON THE COVENTRY CANAL

The south end of the waterway from Hawkesbury into Coventry itself is a dead-end and therefore much less used than the rest of the canal. This, and the fact that it passes through a heavily built-up area, tends to make it a little shallow in places and also a target for the rubbish-dumping brigade. However, this is no reason for not navigating it—you simply need to keep your eyes open and have a good boat hook at the ready. Another notorious scourge of the *Coventry Canal* has always been the slurry pumped into the canal from the quarry concerns in the Hartshill area, although there are signs that this situation is improving. Otherwise, the canal presents no problems to the navigator who will have an easy time of it, except for passing through the Atherstone locks; these always seem to take their time emptying. Boating concerns and facilities are a little thin on the ground, but there is a useful narrowboat specialist firm at Bedworth and good modern boatyards at Nuneaton and Fradley Junction. There are also two large boat clubs at Huddlesford Junction and at Kettlebrook Wharf, near Glascote Locks.

Birmingham and Fazeley Canal

The main link between the *Birmingham Canal Navigation* and the canals to the north and east, the *Birmingham and Fazeley* is historically part of the *Birmingham Canal Navigation*. However, it is hardly a part of the BCN network as such in geographical terms, and is treated separately here.

The canal is a useful but unspectacular Midland waterway, forming a

Farmer's Bridge Flight of locks in the middle of Birmingham, at Brindley House, which is named after the engineer. There is a lock actually beneath the building.

line down from the hills of Birmingham to the industrial lowlands of the Tame Valley near Tamworth. Its characteristics include the faithful number-ing of every lock cottage along the way and the provision in the brick bridges of unusual trap doors set into the arch of the bridge. Behind each of these doors lies a long narrow chamber for the safe and convenient storage of the stop-planks that fit each bridge—unlike the many red-painted trap doors in the parapet of the BCN bridges, which were built to allow firemen access to a large and convenient reservoir of water should they be fighting a city fire nearby. (These trap doors were apparently of great use during air raids in the Second World War.)

The *Birmingham and Fazeley* starts in the centre of Birmingham, runs the whole gamut of that city's urban characteristics and then takes us out into the countryside. This accounts for the top eight miles of the canal in contrast with its open course from Curdworth down to Tamworth. The top end of the canal in Birmingham itself is very interesting. The waterway leaves the main line of the *Birmingham Canal* at Farmers Bridge, opening out into a large basin overlooked by a new canal pub. From here the canal literally disappears down a steep flight of locks through Birmingham. The city crowds it on every side and even from above. (The tall Post Office Tower straddles one of the locks.) The canal remains enclosed by this dark, narrow corridor even after levelling out when the Farmers Bridge locks tail off. At Aston Junction the canal divides, the Bordesley Branch leading off to the south and east to negotiate six descending locks, a narrow tunnel and a cavernous multiple railway bridge before terminating at the old Bordesley Basin. The *Grand Union* joins nearby.

The main line of the *Birmingham and Fazeley,* north-east from Aston Junction, plunges down another flight of locks through the unsunny nether regions of Aston. The canal's course is not quite as claustrophobic as

At the village of Drayton Bassett there is a footbridge 'folly' over the *Birmingham and Fazeley Canal.*

it is in the Farmer's Bridge flight, but it is still very much shut away from the great conurbation on either side.

The *Birmingham and Fazeley*'s junctions at Salford with the *Tame Valley Canal* and the *Saltley Cut*—once a rather bleak spot in the Black Country hinterland—are now entirely covered over by the indelicate tracery of countless, curving elevated roadways comprising the famous motorway interchange known as 'Spaghetti Junction'. The various canals creep between the concrete feet of this gigantic flyover. From here the *Birmingham and Fazeley Canal* continues north-east towards Fazeley; it is an unromantic stretch. The water is dirty and the lock walls usually coated with oil and filth. There is a noisy road alongside, and precious little merit in the scenery save what is to be found in places like 'Fort Dunlop' and the inside of a great warehouse which has been built right over the waterway.

But once down the Minworth locks and past a great sewage works, this locale is changed for a more rural course. Curdworth Tunnel, an attractive but minuscule eighty-one yard affair through solid rock, leads to the isolated locks of the Curdworth flight; there is a good canal-side pub halfway down. North of the locks, the canal adopts a straight course through low-lying and somehow unconvincing open country in which the towpath hedge seems a distinct advantage—although there are also some encouraging groups of young and old oak trees along the way. Near the village of Drayton Bassett the canal is crossed by a 'folly' bridge of tiny, castellated twin white towers joined by a wooden foot bridge. At Fazeley, the *Birmingham and Fazeley* joins the *Coventry Canal* and the former 'owns' the latter's line along the Tame valley past Hopwas Woods as far as Whittington.

PLACES TO SEE THE CANAL

The intensely urban stretch at the Farmer's Bridge end of the canal is a fascinating waterway to explore, although the authorities do their best to prevent the pedestrian from getting on to the towpath. Salford Junction, with the overhead web of motorways, is positively awe-inspiring; but a more relaxing place to see the *Birmingham and Fazeley* would be along Curdworth Locks. A stroll here on a summer's evening is very rewarding, especially if you end your walk at the 'Dog & Doublet' on Bodymoor Heath. (However, it is ominous that the M42 motorway is projected to be built parallel and close to these locks.) The rocky Curdworth Tunnel is appealing, if only for the great cutting at its north end; but beware of the slippery towpath inside. The curious footbridge beside the main road at Drayton Bassett is worth a look, and so are the environs of Fazeley Junction, with its numbered cottages and the nearby canal-side mill.

Ashby Canal

There is little similarity between the *Coventry Canal* and its dependent waterway, the *Ashby,* for although the two canals are never more than seven miles apart, the *Ashby* is just far enough east to avoid the reaches of the Bedworth coalfield, and the surrounding countryside remains unexploited and undespoiled. In addition, there are no connections with any other waterway, so the *Ashby* is only lightly trafficked. It is certainly worthy of note, however, that coal is still occasionally loaded into narrow boats at Gopsall Wharf for transport southwards. There are some attractive villages along the route and the bridges are great unfussy arches, built of stone in the south and red or blue brick in the north. There are a handful of minor aqueducts too, but no cuttings or embankments of note; just one tunnel right up at the northern end. The actual channel of the waterway is unusual, being not only clean and fairly deep but also lined by banks which are entirely unprotected by any form of walling, a useful economy measure in the construction of the canal but something which has its disadvantages in these days of powered craft. Perhaps the weed fringes will continue to suffice to shield the banks from the wash of the boats.

The *Ashby Canal* was originally conceived as a link between the *Coventry Canal* and the River Trent in the days when the latter was navigable right up to Burton-on-Trent. The advent of the *Grand Junction Canal* scheme, and the *Grand Junction's* hope that the Oxford and Coventry companies would widen their locks to extend the *Grand Junction's* broad canal northwards, revived ideas of a broad canal to complete the line to the Trent. The *Ashby Canal* was duly built with wide bridges (there were no locks), but the plan for the Trent link was scrapped in favour of tramways north of Moira, on the Ashby Woulds. The canal therefore extended from Moira to the *Coventry Canal,* and was opened in 1804.

The canal/tramway combination proved successful. The hoped-for development of the coalfields around the Ashby Woulds and Swadlincote area took place, and the success of the canal itself was assured by the discovery of high-quality coal at a new pit sunk at Moira. This coal was thereafter carried away by boat over the full length of the canal to markets all over the south Midlands and the Home Counties. Subsequently, the canal was bought and operated by the Midland Railway whose iron boundary markers can still be seen in fields along the offside of the canal.

The length of the *Ashby Canal* varied wildly in the planning stages. As built, it ran thirty miles from Marston Junction to Moira, but its nickname which still persists—the 'Moira Cut'—is now outdated, for the canal no

The *Ashby Canal* near Marston Junction where it meets the *Coventry Canal*.

longer reaches the town. Subsidence caused by the very collieries it served for so long has seriously undermined the canal bed over the last thirty years, and the waterway has been progressively shortened to a mere twenty-two miles. So what remained of the old tramway system was abandoned, and Moira, Donisthorpe, Oakthorpe and Measham are now places that are cut off from the canal. At the last of these, the dark-brown, ornamented Measham pottery so closely associated with the narrowboat tradition was originated.

The present terminus of the *Ashby Canal.*

The *Ashby Canal* leaves the *Coventry Canal* at Marston Junction on the outskirts of Bedworth (the original junction was to have been near the Griff Colliery's canal outside Nuneaton), and immediately exchanges the mining district for an agricultural neighbourhood in which it remains all the way to its terminus. The canal passes Burton Hastings, the Fosse Way, and the manufacturing town of Hinckley where there is a tiny branch, or large dock, towards the town. North of Hinckley, the canal is joined by an old railway which traces a rather straightened version of the canal's own course past the villages of Stoke Golding, Shenton, Congerstone, Shackerstone and right up to Snarestone. At Snarestone, a shallow 250 yard tunnel takes the canal under the village street, and the canal's new terminus is only another three quarters of a mile further on, near an isolated Victorian waterworks.

The canal, passing through a very quiet and unknown tract of Warwickshire and Leicestershire, is more notable for its quiet rural environment than for spectacular engineering achievements. There is a heavy, single-arched aqueduct over a minor road at Shenton, and the tunnel at Snarestone where the towpath's special lane over the hill can still be traced. The canal's only lock, the stop lock at Marston Junction, is now unused although it remains intact. Like the bridges on the canal it was built wide enough for barge traffic but the lock was later narrowed.

Non-canal curiosities along the waterway include a 'live' steam railway museum at the old Shackerstone Station and, near Shenton, the site of the Battle of Bosworth where Henry Tudor's victory over King Richard III in 1485 signalled the end of the Wars of the Roses and the beginning of the Tudor dynasty. Henry became King Henry VII and first donned the crown just after the battle, on a hill at Stoke Golding.

43

The *Ashby* is ideal boating water for those who are not mad about locks, for there is not one throughout its length, except for the old stop lock at the south end. The nature of the surrounding countryside also enhances the canal, and the depth of the water is helpful, while the great width and height of the bridgeholes—a reminder of the high hopes for this canal as part of a Thames to Trent barge link—is a note of sheer luxury. But a slight drawback is that the 'soft' edges of the waterway make mooring rather difficult: a gangplank can be very useful on this canal. There is not an excess of winding holes along the route, but they now include one at the very end, north of Snarestone. (There is also a slipway and a car park at the terminus.) Small pleasure boats need to steer well clear of any commercial narrowboats under way, especially if they are loaded, because their passage can cause quite a rush of water, particularly in shallow sections.

2

Birmingham Canal Navigations

BIRMINGHAM CANAL
WYRLEY AND ESSINGTON CANAL
TAME VALLEY CANAL
WALSALL CANAL
TITFORD CANAL
DUDLEY CANAL
STOURBRIDGE CANAL

A network within a network, the waterways of the *Birmingham Canal Navigations* are even today impressive as a comprehensive transport system for a huge manufacturing and mining area. The fact that they are today virtually silent and unused, jostled on all sides by the roads and railways that were their undoing, makes them all the more fascinating.

The *Birmingham Canal Navigations,* or BCN as they are usually called, comprise that tangle of narrow waterways that thread their way through the Black Country. These canals, which were among the earliest in the Midlands, were more than any other single factor instrumental in the very fast development of small industrial towns like Birmingham, Walsall, Dudley and Wolverhampton, and of villages like Tipton, Bilston and Oldbury. The canals linked all these places, joining not only factory to factory but factory to colliery too, for the carriage of fuel for the Industrial Revolution. (Later, canals were extensively used to supply coal to the several power stations built along the BCN.) This role has now entirely disappeared, except for a minimal amount of goods carried for short distances here and there by day-boats, which are mostly cabin-less, engine-less narrow boats towed by a tug or—even today—a horse.

On other canals, the decline of commercial traffic has coincided with the boom in pleasure boating, an activity for which most of the waterways are ideally suited. But conditions are somewhat different on the BCN. The general surroundings of a built-up, industrial area through whose oldest and most decrepit parts many of these canals seem to run, and the hazards of navigating canals which frequently suffer from vandalism and the dumping of rubbish, filthy water and even filthier locks, are all conditions which tend, understandably, to deter the average boater. In addition, the lack of boatyards and other facilities, and the physical isolation of the canals through much of the network, keeps the canals of the Black Country in a strange sort of limbo. Many canalside pubs, for example, are isolated by high walls from the canals they used to serve. But the *Birmingham Canal Navigations,* which in their extent and character are entirely unique in this country, make up a group of waterways well worthy of exploration.

Canals in the Black Country originated in Birmingham, with a scheme floated in 1767 to build a canal to Wolverhampton, taking in the coalfields between the two towns and joining the new *Staffordshire and Worcestershire Canal* beyond Wolverhampton. The canal was to link Birmingham not only to the Severn but also to the river basins being linked by the *Trent and Mersey Canal.* This winding waterway was duly built. Called the *Birmingham Canal,* it formed the axis of all the branches and other canals subsequently built. These connecting waterways soon formed a system with Birmingham at its south-east corner, but the *Birmingham Canal* remained at all times the principal canal and the principal company, taking over or merging with all the others.

The system has been heavily pruned since its heyday a century ago, and the navigable network that remains extends to only a hundred miles or so. Fortunately, however, most of the principal through-routes remain, and an exploration of the BCN is a fascinating and unique experience. Cuttings and embankments, tunnels, canal flyovers, obscure branches and long flights of locks are here in abundance. The flyovers themselves are of unique importance in the history of civil engineering; two of the tunnels (at Dudley

BCN—Stewart Aqueduct, carrying one canal over another.

and Netherton) are, respectively, the longest and widest navigable canal tunnels in Britain; and the waterways themselves offer an unrivalled back-door glimpse not only of the heart of the Industrial Revolution but also of modern manufacturing establishments. And while some of the names connected with the BCN seem unlikely (Pudding Green Junction, Gosty Hill Tunnel, Oozells Street Loop, Sneyd Bank, Titford Pools, and even the *Wyrley and Essington Canal* itself), others like Bilston, Soho, Horseley, Smethwick and Walsall are historic industrial place names whose eminence is directly attributable to the cheap and convenient transport of fuel, raw materials and finished products provided by the canals of the BCN.

The extensive construction of motorways and other primary roads around the Black Country has, in the last ten years, had a dramatic effect on some of the canals. Some, perhaps inevitably, were closed in the course of the road construction programme, but others now find themselves roofed over by giant concrete decks and flanked by lines of huge concrete columns. The adjacency of the ancient and modern transport routes can seem almost surreal. At Spon Lane junction, for example, a historic canal aqueduct carries one waterway over another a few yards from the three Spon Lane locks—believed to be the oldest working locks in the country. Right overhead is the M5 motorway; and beside the aqueduct is an entirely up-to-date electrified railway. Elsewhere, the M5 has occasioned the building of a large new aqueduct to carry the *Tame Valley Canal* over the road. The east end of the same canal is almost smothered by the futuristic curves of the 'Spaghetti Junction' motorway interchange; while in Smethwick, a pair of new concrete tunnels have been built to carry the canal beneath a large new dual carriageway. Such modern encroachments, which would be so grotesque on more rural waterways, seem very much in keeping with the BCN. The effect of the new motorways on the urban scene—and on commercial prosperity—must be very similar today to the effect of the coming of the

New motorways in and around Birmingham have provided lengths of canal with useful shelter from the rain.

Part of the *Birmingham Canal Navigations*.

canals two hundred years ago. Perhaps 'Spaghetti Junction' is the logical descendant of Stewart's Aqueduct?

The map (page 252) gives some idea of the extent of the navigable parts of the *Birmingham Canal Navigations,* but it cannot, of course, show the countless private docks, basins and arms which used to jut out along the course of so many of the waterways. These are now almost all unused, blocked either by shallow water or by permanent infilling. Nor can the map show the extent of the industry that was once so dependent upon the waterways; and there is no room here even to touch on the subject of the reservoirs and the complicated system of pumping engines—originally all steam-powered and housed in great two- or three-storey buildings scattered about the Black Country—that were built to maintain the water levels on the heavily used canals of the BCN. (These pumping engines deserve a separate study of their own.) But the map does provide the basic data to show where boats can travel in an unbroken line, and all the locks that intervene. The latter are nearly all grouped in flights, and are thus easily identified. The principal components of the BCN are briefly summarized below.

Birmingham Canal

The *Birmingham Canal,* the spine of the system, probably best expresses the overall 'feel' of the BCN today. The canal is more than a single line, for the original circuitous route was shortened drastically by Thomas Telford in the 1830s to make a quite separate waterway for much of the distance. Fortunately, most of the original canal survives. Hence the 'loops' at the south-east end, the parallel but very different waterways through Smethwick, Dudley and Tipton, and the canal aqueducts en route. Indeed, these few miles contain many of the most historic structures on the BCN. The three miles of dead straight waterway (the 'new' main line engineered by Telford) between Oldbury and Tipton must qualify as one of the bleakest on the network, relieved only by the fine view of the bulbous brown ridge, to the west, that is penetrated by the Dudley and Netherton Tunnels. Further north, the old and new lines join to pass through Coseley Tunnel which, with its large cross-section and twin towpaths, smacks of Newbold Tunnel on the *Oxford Canal.* Both were engineered by Telford; Coseley cuts out a long deviation of which the Wednesbury Oak Loop—a sinuous two mile branch now leading only to some lock-gate workshops—is the truncated survivor.

49

From Coseley the canal passes the Bilston Steelworks and slips into the industrial back regions of Wolverhampton. A flight of twenty-one locks leads down to the *Staffordshire and Worcestershire Canal* and out of the Black Country.

Wyrley and Essington Canal

A tortuous canal starting near Wolverhampton, the *Wyrley and Essington* once went right through to Huddlesford Junction, on the *Coventry Canal* near Fradley. But today the thirty Ogley locks in the canal's last seven miles are abandoned. The sixteen lock-free miles that remain pass mainly housing estates and would-be countryside on the northern fringes of the Black Country. The canal's slavish adherence to the contours of the land makes it extremely difficult for steerers of full length boats to avoid minor collisions with other craft. One of the worst of the corners is at Sneyd Junction from where a branch used to continue north to Cheslyn Hay. Other notable offshoots are the remains of the *Bentley Canal* at Wednesfield Junction; the *Cannock Extension Canal,* a straight and late (1850s) waterway now cut short by the A5 but sporting a huddle of three of the only boatyards on the BCN; the Anglesey Branch, which leads up to the head bank of the great Chasewater reservoir; and the Daw End Branch, which perpetuates the *Wyrley and Essington* contour style. Subsidence has forced the banks to be continually raised on this branch, leaving a very deep waterway with rather low bridges. At Longwood a lock marks the junction with the much younger *Rushall Canal*—an unfussy, 'non-contour' but pleasantly rural link via nine locks with the *Tame Valley Canal.*

Tame Valley Canal

The *Tame Valley* was built late in the day, designed as a bypass to the vulnerable bottleneck of the Farmer's Bridge locks in Birmingham. Opened only in 1844, the canal is a great contrast to most other BCN waterways, its straight course leading it through long steep cuttings and over tall embank-ments. There are several aqueducts, most of which are insignificant; but the new one over the M5 motorway is very impressive. The canal flirts with the M6 for much of the way, especially towards the bottom end of the thirteen Perry Barr locks which link it to the *Birmingham and Fazeley*.

Walsall Canal

An aggressively industrial waterway taking in all the extremes of the BCN, the canal leaves the *Wednesbury Old Canal* at Riders Green and drops down a corridor of eight locks before proceeding on the level round to Walsall, where a one mile branch leads up to the deep Walsall locks—eight of them—and back to the 'Wolverhampton Level' of the *Wyrley and Essington Canal*. A veritable tangle of minor canals used to join the *Walsall Canal* in the Ocker Hill/Wednesbury area; but now such names as the Danks Branch, the Gospel Oak Branch and the Tipton Green and Toll End Communication no longer have currency. Only part of the Ridgacre Branch survives as a spur of the *Wednesbury Old Canal*—a line where working boats are still found in action—and further towards Walsall, the Anson Branch is still theoretically navigable, although shallow. This branch used to form part of the *Bentley Canal's* through route up to the *Wyrley and Essington*.

Titford Canal

Six locks from the *Birmingham Canal's* old main line at Oldbury, starting in the lee of the M5 motorway, lift boats up to the highest level on the BCN, 511 feet above sea level. The terminus is now Titford Pool; and the whole canal to this point has recently been restored to full navigability.

Dudley Canal

The *Dudley Canal* was drawn into the BCN net at a fairly late stage, having been a fierce rival of the Birmingham Canal Company for many years. The *Dudley Canal* itself is an almost self-contained system, especially if one

Langley Maltings on the *Titford Canal* (BCN).

includes its western neighbour, the *Stourbridge Canal*. (The *Dudley's* principal *raison d'être* was to supply coal to the glassworks for which Stourbridge is famous even today.) The principal features are the two long tunnels that cut through the great ridge, astride which sits the town of Dudley, overlooking miles of the Black Country. Although similar in length, the tunnels could hardly be more different in character. The Dudley is perhaps the most complicated canal tunnel in the country (apart from the network at Worsley, on the *Bridgewater Canal*), for it was built to provide access to collieries and limestone quarries right inside the hill. There are branches and old workings within the tunnel itself, and large caverns and basins which open to the air far above. Much of the tunnel is cut through solid rock and unlined, and its dimensions are claustrophobic in the extreme. There is room for only one low-slung boat at a time, and as there is no towpath, poling or legging was the order of the day. Netherton Tunnel runs parallel to it and was built to relieve congestion in the Dudley Tunnel. It was one of the last canal tunnels to be built and, by contrast, is wide and straight. It also has the unique luxury of twin towpaths and was originally lit by gas along all of its impressive 3027 yards.

The *Dudley Canal* was originally a short waterway, joining the *Stourbridge Canal* at Delph to give access to the coal mines around Netherton, south of Dudley. A branch was soon built as a tunnel to reach the separate mines and quarries already being worked inside Dudley Hill, and this extension gave access to the *Birmingham Canal* at Tipton Hill. Meanwhile, a second line was built from the Netherton terminus to go eastwards and round to the *Worcester and Birmingham Canal* at Selly Oak, following the contours of the land to begin with, and then taking in two tunnels—one at Gosty Hill

The start of the Netherton Tunnel on the *Dudley Canal,* built to relieve the congestion on and parallel to the Dudley Tunnel. You can see the unusual twin towpaths.

The approach to the Netherton Tunnel—the opening is just visible under the brick bridge by the derelict pumping station. The elegant iron footbridges are very characteristic of the BCN.

and a very long one (3795 yards) at Lappal. The latter, little more than a large drainpipe, is now closed, leaving Coombeswood Basin (just south of Gosty Hill tunnel) as the terminus. Although it seems a little out on a BCN limb, Coombeswood remains an active pocket of canal life, with narrowboats still used for the storage and transport of steel components within the works.

The other point of particular interest on the *Dudley Canal* is the flight of eight locks at Delph, dating from 1858, at Brierley Hill at the *Stourbridge Canal* end. Previously, there was a flight of nine locks whose remains can still be seen to one side.

BOATING ON THE BIRMINGHAM CANAL NAVIGATIONS

It should already be evident what to expect on a boating trip on the BCN; the problems are principally those of most narrow, industrial canals. Vandalism and lack of maintenance often render lock operation difficult, and it is not uncommon to find short pounds dry. Rubbish in the channel ranges from the usual propellor-fouling plastic bits and pieces and lengths of old rope, to submerged oil drums and more dangerous obstructions like steel pipework and other industrial debris. Many of the old canal arms and loops tend to be at least partially blocked by such things, and fibre-glass

The Black Delph Flight at Brierley Hill on the *Dudley Canal* (BCN). The open cut mine stands
square at the end of the run.

boats are especially at risk if a sharp lookout is not kept on the water ahead. Great agility is also necessary, for high brick walls often cut off the more urban of the canals from their surroundings and thus make it difficult to fetch provisions, visit pubs and so on; boating facilities such as water taps and dustbins are few and far between. One particularly important point on the *Dudley Canal* is that Dudley Tunnel is extremely low, having been designed for dayboats that had no cabins. Boats that can get in are forbidden to use their engines in the tunnel because of the danger from fumes, so poling or legging is the order of the day.

But other factors more than offset these drawbacks; the character of the BCN is by no means a uniformly industrial one. Towards the northern and eastern limbs of the system, reaches are positively rural, an impression enhanced by the cleanness of the water that feeds down from Chasewater reservoir; and because of their extensive use until recent times, most of the canals are still reasonably deep. At any time of year there is an astonishing absence of other boats on the move, a factor which, with the decay all around, makes a trip along parts of the *Birmingham Canal Navigations* seem in many ways like a ride through a ghost town. Great basins lie abandoned and empty of all but the occasional hulks of one-time working boats; old toll houses stand derelict, and factories and works everywhere have turned their backs to the canal. The private world of the BCN seems empty even of people: few lock keepers and fewer boatmen are in evidence.

Two of the system's focal points, where people and boats congregate in any number, are at the new development at Birmingham's Cambrian Wharf on the *Birmingham and Fazeley Canal* (described elsewhere although historically part of the BCN), and at Gas Street Basin, an extraordinary and very

Gas Street Basin, Birmingham.

much alive canal community tucked away in Birmingham itself. Rows of moored narrowboats offer here a single glimpse of the good old days on the BCN, giving also the impression of a united front against long-standing plans drastically to modernize and open out this hidden backwater.

Two more isolated, and perhaps more realistic, centres of activity on the BCN are the group of traditional boatyards at Norton Canes on the Cannock Extension, and Alfred Matty's yard at Coseley, from which carrying is still undertaken. Other BCN communities have just faded away.

Stourbridge Canal

Historically speaking, the *Stourbridge Canal* was never a part of the *Birmingham Canal Navigations* and remained an independent—and profitable—company up to the time of its nationalization. It is, however, a short canal, almost entirely within the confines of the Black Country, and it has always been closely connected to the *Dudley Canal.*

The *Stourbridge Canal,* therefore, is appended here to the canals of the BCN. Along with the *Dudley Canal,* this waterway was built to fetch coal from the mines around Dudley to industry at Stourbridge—principally the already well-established glass industry—in addition to providing the town with an outlet to the *Staffordshire and Worcestershire Canal* and thus to the Severn. The canal is still a most useful link route and is, despite its mere five and a quarter miles, a canal of dramatic contrasts and considerable interest, especially in the sphere of industrial history.

The main line of the *Stourbridge Canal* today is the route from Stourton Junction on the *Staffs and Worcs Canal* to the meeting with the *Dudley* at the foot of Delph Locks, the line into Stourbridge itself being considered as the 'Stourbridge Branch'. The old lines to the reservoirs of Pensnett Chase, the *Stourbridge Extension Canal* and most of the Stourbridge Branch itself are now all unnavigable. (There is a plan to reopen at least a substantial part of the latter.)

This short canal is notable mainly for the steep flight of locks—the 'Stourbridge Sixteen'—which carries it up past the glassworks and on to a level pound that doubles back along the hillside before twisting through an intensely industrial area to the *Dudley Canal* at Delph Locks. The sixteen locks themselves are a fortunate survival, having been rescued from derelic-tion in the nick of time and restored mainly by the sweat of voluntary labour between 1964 and 1967.

In the 'Stourbridge Sixteen' locks.

The lower, western end of the canal is quite different, comprising a pleasantly wooded stretch that seems utterly removed from the industry of Stourbridge or Brierley Hill. Four locks lead the canal down to a tree-shaded junction with the rural *Staffs and Worcs Canal* just 'upstream' of Stewponey Lock.

3

Grand Union Canal System

The *Grand Union* is not a single canal but a system of waterways connecting London and the Home Counties with the West and East Midlands. The present *Grand Union Canal* was formed as late as 1929 by the amalgamation of several linked waterways. Other canals were later taken over by the *Grand Union,* and their geographical position has tended to somewhat blur the edges of the *Grand Union* umbrella. Thus, although the *Grand Union* today is normally taken as being the canal from Paddington and Brentford (its junction with the River Thames) to Birmingham—with a major branch to Leicester and minor branches to Slough, Aylesbury, Northampton, Welford and Market Harborough—in fact, the canal technically includes the River Soar from Leicester to the Trent, and the quite separate *Erewash Canal*—in addition to the whole of the *Regent's Canal* in London (with its connecting link, the *Hertford Union*) and the *Saltley Cut* near Birmingham.

One may further confuse the situation by pointing to the course shared by the *Grand Union* and the *Oxford* between Braunston and Napton, and to the existence of a previous (and much less grand) *Grand Union* which is now but a tiny section of the present *Grand Union* system. It would perhaps be easiest to unravel this geographical and historical tangle by tracing, at least in outline, the present *Grand Union*'s history from the start.

It was the *Oxford Canal,* opened throughout in 1789, which first allowed the possibility of traffic between the capital, the industrial Midlands and the north-east and north-west (via the *Trent and Mersey Canal*). It was a boon indeed, but it was a slow, winding narrow route which relied on the River Thames with all the difficulties of haulage and flooding that were associated with a river navigation. Sure enough, a better route was soon planned which envisaged a broad canal from halfway up the *Oxford Canal* to join the tideway of the River Thames at Brentford, west of London. This was named the *Grand Junction Canal*. Its short route and broad locks must have looked threatening to the owners of the now-outdated *Oxford Canal,* and the latter had to be appeased by guarantees of substantial minimum toll revenues: the *Grand Junction* as built (it was completed throughout in 1805) ran for ninety-three and three quarter miles, from Brentford on the Thames to Braunston on the *Oxford Canal*. From there, Birmingham-bound traffic used the *Oxford, Coventry* and *Birmingham and Fazeley* canals, but that was not the end of the story. The coming of the great new waterway occasioned other schemes, and the *Warwick and Napton* and *Warwick and Birmingham* canals were soon projected as a new short cut to Birmingham. The former's junction with the *Oxford Canal* at Napton helped to divide the *Oxford Canal* into northern and southern sections, separated by five miles of waterway that were now shared with the *Grand Junction*. Branches were built, one of the most important being the line into London, a lockless branch of thirteen and a half miles from Bulls Bridge (near Southall) into the heart of Paddington. This was opened in 1801, and later spawned the *Regent's*

Braunston Bottom Lock on the *Grand Union Canal.* The building on the left is a covered dock.

Hanwell Locks on the *Grand Union* near Brentford.

Canal, a waterway which circled round central London and dropped through a dozen paired locks to a dock off the Thames at Limehouse. This was opened in 1820.

Meanwhile, another important development was taking place towards the north end of the *Grand Junction Canal*. In 1814, the original *Grand Union* was opened from Norton Junction on the *Grand Junction* to Foxton, linking up with the *Leicestershire and Northamptonshire Canal* which had started well at Leicester and stalled just short of Foxton, fully seventeen years earlier. Closing this gap at last created a direct through route from the *Grand Junction* to Leicestershire, Nottinghamshire and Derbyshire. The benefits of having several interdependent canals under one 'roof' were obvious to the *Grand Junction Canal*'s directors for many years, but it was not until 1894 that the company managed to acquire the old *Grand Union* and the *Leicestershire and*

Northamptonshire Union, the two struggling canals that between them owned the line down to Leicester and the River Soar.

Further moves towards a united canal came in the 1920s, but this time it was the *Regent's Canal* that made the running, taking over the *Grand Junction* group and the two narrow *Warwick* canals in the late 1920s. The new *Grand Union Canal* itself came into being in 1929 and shortly afterwards absorbed the *Leicester Canal* (the Soar valley from Leicester to Loughborough), the *Loughborough Navigation* from Loughborough to the Trent, and even the *Erewash Canal* on the other side of the Trent.

Grand Junction Canal

This is the spine of the *Grand Union* system. The *Grand Junction* was built as a brave new venture, a wide waterway penetrating from the Thames Basin to the heart of narrowboat territory in the Midlands. Born in the hope of achieving the same sort of importance as the *Trent and Mersey* some twenty-five years earlier, the *Grand Junction Canal's* opening in 1805 represented the same scale of advance in transport thinking as the M1 motorway in 1959— and along the same route. Although the point about having wide locks to allow bigger boats to carry more cargo was lost on the neighbouring canal companies (upon which the *Grand Junction* was relying to further its own wide waterway line), the latter was nonetheless a great success, and usually in the forefront of progress.

The route followed by the *Grand Junction,* taking in both the Northamptonshire Heights and the Chiltern Hills (separated as they are by the valley of the Ouse and flanked by the valleys of the Warwickshire Avon and the Thames), was a route bound to involve heavy lockage; and indeed, despite the two long tunnels at Blisworth and Braunston and the very long cutting along the Tring summit level, no less than 110 locks were necessary, although nine of these at the crossing of the River Ouse near Wolverton were replaced shortly afterwards by an embankment and an aqueduct over the river.

The canal is one of wide open views, endless locks, and plenty of important branches. It is also a very attractive waterway, which passes through virtually no industrial areas apart from its southern end. Instead, it seems to enjoy the best parts of the Home Counties.

The *Grand Junction* leaves the *Oxford Canal* at Braunston, a strategic waterway junction since 1800. Then the six Braunston Locks take the canal

The iron weighbridge at Stoke Bruerne on the *Grand Union*.

to the mouth of the 2042 yard Braunston Tunnel. At the other end of this is an undulating and lightly wooded countryside. Up and to the south is Daventry and the reservoir that feeds this summit level. The other source of water is the canal's Leicester section, which enters by the swing bridge and pretty cottage at Norton Junction. Nearby, the main line ducks under the

A5 road at Buckby Top Lock; most of the remaining locks in the flight are hedged in by the M1 motorway and the London–Birmingham railway line. Fortunately for canal users, the motorway does not stay beside the canal for long, although the railway follows the approximate route of the canal all the way back to London. The railway and canal traverse the village of Weedon together, the canal being carried across a pair of small aqueducts past the church below. In this unlikely spot is a barracks dating back to the days of George the Third; the canal branch (now closed) that served the barracks carried the title of 'The Weedon Military Dock'.

From the new marina at Buckby Bottom Lock to Stoke Bruerne, the canal holds level for fifteen miles at about the 300 foot contour. It gives Northampton a wide berth but is connected with it via the Northampton Arm. The main line of the canal continues almost lock-free for twenty miles. This stretch takes it through the rather bland valley of the River Ouse and then out of Northamptonshire into North Buckinghamshire. It passes the railway town of Wolverton where the great carriage works are right beside the canal. A more drastic and much less warranted blot on the landscape is the new 'city' of Milton Keynes, which has devastated a once-pretty valley (the Ouzel) and ruined a string of once charming villages. The lock up at Fenny Stratford hints at the long haul to the south over the Chiltern Hills. By Leighton Buzzard, the hills are clearly in view and the locks become ever steeper. The double-arched bridges over the foot of some of these locks show that they were once scheduled to be paired locks.

There are several good examples of the early and very decorative style of *Grand Junction* architecture in the lockside buildings, with their elegant fanlight windows. Indeed, even without the Chiltern overtones there are

An ornamental bridge in Cassiobury Park, near Watford on the *Grand Union Canal*.

Knowle Flight on the *Warwick and Birmingham,* about ten miles from Birmingham. Note the side pounds, to save water.

some delightful spots along the canal, like the isolated little church at Church Lock, and the lockside farm near Slapton. Where the canal reaches up to Marsworth, a steep flight of narrow locks drops westward down to Aylesbury—a worthwhile diversion even if it is a cul de sac. The branch is just over six miles long. At the top of yet another group of seven broad locks is the summit level, fed from the adjacent reservoirs via what remains of the Wendover Arm. The Tring summit is a cutting three miles long, at the end of which the Cow Roast Lock begins the long descent to the Thames. A constant succession of locks takes the waterway steadily down from here through the built-up but (from the canal) attractive area. The various settlements along the tree-dotted valley present the canal with a good mixture of mills, cottages, trees, lawns and pubs. The small River Gade runs parallel to the canal.

At Watford the canal avoids the town and passes, instead, through the beautiful surroundings of Cassiobury Park, which are all the better for an

Hatton Flight on the *Warwick and Birmingham* with a panoramic view back over Warwick itself. The protective paddle gearing caps are distinctive on this canal.

ornamental, balustraded bridge. Below Watford, the River Gade is joined by the rivers Colne and Chess, and their combined valley becomes a disjointed series of flooded gravel pits. At Cowley, near Uxbridge, comes one of the last locks before London; but before the Thames is reached, the canal negotiates a remarkably drab and anonymous area on the western outskirts of London. There is a flight of locks at Hanwell and a few isolated stragglers before Brentford. Rather unprepossessingly tucked away at the back of Brentford, the *Grand Union* finally drops into a backwater of the tidal Thames opposite Kew Gardens.

THE BRAUNSTON TO BIRMINGHAM SECTION

The *Grand Junction*'s north-western terminus was a junction with the *Oxford Canal* at Braunston. For a short while after the former's opening, traffic for the important coal mining and industrial area of the Black Country had to

66

go north up the *Oxford* and *Coventry* canals, then back up the *Birmingham and Fazeley Canal,* through Salford and Aston Junctions to Farmer's Bridge (Birmingham). This was a long route along narrow canals which were already dating fast—especially the northern half of the *Oxford Canal,* with its circuitous course.

The *Warwick and Birmingham* and *Warwick and Napton* canals were built partly in anticipation of this pointless diversion, by forming a single water-way from Birmingham to Napton, on the *Oxford Canal.* By then using the *Oxford Canal* between Braunston and Napton, the *Grand Junction* route was extended right through from the Thames to the *Birmingham Canal Navigations.* This became the main line of the *Grand Union Canal* system, and when the big modernization scheme was instituted in the 1930s, it was

Navigation Inn on the *Soar Navigation,* at Barrow-on-Soar.

the two *Warwick* canals with their long flights of narrow locks upon which most attention was necessarily focused. The narrow locks had been installed on the *Warwick* canals to the obvious dismay of the Grand Junction Canal Company, who saw the chance of achieving any form of wide waterway north of Braunston suffer yet another setback. The increased convenience and carrying potential of a wide waterway was no doubt more than outweighed by not only the increased construction cost of such locks, but also the much greater water supply required to fill the big chambers, and an understandable reluctance to vary from the seven feet beam standard adhered to by the canals of the West Midlands.

The rebuilding of the canals to a potential barge standard, with the reconstruction of all the locks between Napton and Knowle (fifty-two narrow locks were replaced by fifty-one broad ones), seems now to have been an investment which has never really paid off. Widened locks enabled the canal to carry more, but although wide boats—or breasted-up pairs of narrowboats—could theoretically penetrate from London and the Thames to the edge of Birmingham, the awesome problem of widening the many flights of locks picking their way through Birmingham was never tackled. Indeed in many places, the channel of the waterway was never made sufficiently wide or deep to pass two wide boats—not to mention the tunnels en route. And lastly, of course, the improved waterway did nothing to improve journey times—and hence the cost—of the carriage of goods. The immediate effect of the improvement scheme was to boost carrying on the *Grand Union Canal* main line, but in the long run it could only prolong the end of the canal's commercial life—an end which has become a reality only in the last few years.

THE COURSE OF THE GRAND UNION

BRAUNSTON TO BIRMINGHAM: From Braunston to Napton, the *Grand Union* uses the *Oxford Canal,* a convenience for which the traders had to pay dearly to compensate the old *Oxford Canal* for traffic on both its northern and southern sections lost to the more direct London–Midlands route that the *Grand Junction* and the *Warwick* canals presented.

At Napton Junction the *Warwick and Napton Canal* heads north before falling west through the three Calcutt locks—the first of the 'new' wide locks—passing a series of canal-side pubs around Stockton, where the descent towards the Avon valley starts in earnest. Huge iron paddle posts enclose very low-geared paddles which open large sluices in the locks; all these big locks are very fast to fill and empty. The locks continue to fall through quiet sheltered countryside to Leamington Spa. At some of them, the old narrow

A rare sight nowadays—a narrow boat carrying coal along the *Grand Union Canal*.

locks may be seen alongside. The canal runs straight through Leamington, crosses the Avon on an aqueduct near Warwick, and starts the rise towards Birmingham. The old *Warwick and Napton Canal* ends clearly enough at a T-junction.

The end of the *Warwick and Birmingham Canal* is now called the 'Saltisford Arm', and is derelict. Immediately west is Hatton, an exhausting flight of twenty-one locks with a superb view back over Warwick. Beside the canal is the Warwick–Birmingham railway line which struggles up Hatton Bank. West of Hatton on an eight mile pound is a quarter-mile tunnel. Here the canal turns north, winding through hilly countryside past the *Stratford-upon-Avon Canal* and up to a further flight of five deep locks at Knowle. A final ten mile pound takes the waterway along through Solihull and mile upon mile of Birmingham's bleakest nether regions—not dramatically ugly nor fascinatingly industrial, just anonymous, scruffy and dull. At Bordesley, the canal dives down six narrow locks before joining the BCN at Digbeth. A branch from Bordesley Junction leads down the two and a quarter miles (and six locks) of the confusingly-named *Birmingham and Warwick Junction Canal* (usually called the *Saltley Cut*) to meet the *Birmingham and Fazeley* at Salford Junction.

THE 'LEICESTER LINE'—NORTON JUNCTION TO THE RIVER TRENT: The origins of the route lie in the River Soar, a tributary of the River Trent which was made navigable from its junction with the Trent up to Loughborough (a distance of nine and a quarter miles) by an Act of 1766. It was a somewhat crude undertaking, using (by then outmoded) flash weirs. Most of the navigation was the river channel, but the last mile and a half was an artificial cut to the heart of the town. The navigation gave goods from Loughborough an outlet to the Trent, and enabled coal to be brought cheaply south from Derbyshire. Later (1794), the line was extended up the Soar valley to Leicester, this time using more canal cuts where the river was circuitous or shallow, and dispensing with the use of flash weirs. Two rather short-lived offshoots from this navigation were the Charnwood Forest line (a combined canal/tramway leading west from Loughborough Wharf), the canalization of the River Wreake north of Leicester, and a canal extending the latter navigation from Melton Mowbray up to Oakham.

South of Leicester, the proposed construction of the *Grand Junction Canal* from the Thames to Braunston prompted ideas of connecting the *Soar Navigation* to the *Grand Junction,* thus forming a wide Thames–Trent water link: this was the origin of the *Leicester and Northamptonshire Union Canal* which was actually authorized by Act of Parliament in 1793 as a long winding route south-east from Leicester to the River Nene at Northampton, where it was to join a branch of the new *Grand Junction Canal*. The *Leicestershire and Northamptonshire Union Canal* (generally known as the *Old Union Canal*) was never completed to Northampton, since the canal's builders had run out of money by the time they reached Market Harborough. The rest of the line was abandoned and, after intervention by the Grand Junction Canal Company, the 'Grand Union' was promoted in 1810 to forge the final link in the chain. The canal was completed in 1814, thus opening the Leicester line throughout.

The decision to build a narrow staircase of locks on the *Grand Union Canal* at both Foxton and Watford was a fateful one, ensuring that each place became a bottleneck, limiting the beam of boats, and causing time-wasting hold-ups for such boats as could get through. The opening of the Foxton Inclined Plane in 1900 solved both these drawbacks for a while, but without the widening of Watford Locks or the installation of some similar lift in their place, the Foxton plane was only half useful. After the closure of the latter in 1910, the bottleneck was restored, and it remains as notorious as ever to this day. It is not entirely surprising that through-traffic on the Leicester Line dropped quickly away in the present century.

THE LEICESTER SECTION: The historical background of what is now the *Grand Union*'s Leicester Section accords well with the geography and can be fairly divided into three separate lengths of waterway: the narrow canal from Norton Junction to Foxton, the wide canal from Market Harborough via Foxton to south of Leicester, and the *Soar Navigation* from Leicester to the Trent.

Of these three, the canal between Norton and Foxton junctions is the most remote. It is long and winding, an utterly isolated waterway which ambles along for miles through beautiful upland countryside. Each end of the canal is reached by a short, sharp flight of locks—ten at Foxton and seven at Watford; these locks enclose a level pound twenty-one miles long. A short arm to the village of Welford acts as a water-supply canal from a group of reservoirs at the headwaters of the River Avon. This water flows out of the long summit at either end, a vital source for both the Market Harborough Arm and, at the south end, the Braunston Summit on the *Grand Union*'s main line. However, the long Leicester summit is of great value in its own right, for its sleepy meandering along the contours more than 400 feet above sea level makes it not just the highest canal in the East Midlands but also one of the most delightfully quiet canals in the whole country. There are no towns en route, and any villages are at arm's length from the canal. North of the M1 crossing at the head of Watford Locks, only three

71

classified roads cross the canal, one of them at Husbands Bosworth where the canal is in a tunnel anyway. There is another tunnel at Crick; both add to the sense of rural seclusion that shrouds this lonely waterway. Not surprisingly, it is excellent territory for the naturalist. The northern limit of the summit level ends abruptly at Foxton with a staircase of ten locks down to the Market Harborough level.

The canal from Market Harborough to Leicester is also very attractive, although its remoteness is somewhat compromised by a busy railway line that is never far away. The Market Harborough Arm, from the town to Foxton, is rather narrow and very winding with only the village of Foxton nearby; but north of this, the canal is of more generous dimensions, although it continues to follow an irregular course as it skirts a range of hills to the west. There is a tunnel near Smeeton Westerby, and then the first of the wide locks at Kibworth where the canal drops steadily into the more populated lowlands south of Leicester. At Kings Lock, Aylestone, it joins the River Soar.

From Aylestone to Leicester, Loughborough and the Trent, the *Grand Union Canal* is thus very much a river navigation, in spite of the several long and artificial cuts along the way. Apart from these cuts there are, indeed, few features reminiscent of a canal. The valley is a verdant line of lush water meadows and grazing pasture, with the hills of Charnwood Forest on one side. It is a popular boating and holiday area, with plenty of riverside camping and picnicking possibilities, and crowded riverside pubs. Handsome stone church spires and the occasional water mill further enhance the scene.

THE NORTHAMPTON ARM: A five mile arm from the *Grand Junction's* main line near Blisworth to the River Nene in Northampton, this canal's significance is in providing the only water link between the River Nene and the Fenland waterways on the one hand, and the *Grand Junction* and the rest of the canal system on the other. Unfortunately the narrow gauge of the seventeen locks on the Arm prevents the link being used by wide craft. Those boats that do manage to squeeze in will find an open, rural canal dropping steeply down to the valley of the Nene, with a short, wide, and echoing concrete tunnel halfway down which carries the M1 motorway. A curiosity on the Arm are the small bascule bridges among the locks; these would look more at home on the distant *Llangollen Canal*.

THE AYLESBURY ARM: A delightful canal, six miles long, which falls through sixteen narrow locks from the *Grand Union* main line at Marsworth almost to the centre of Aylesbury. The Aylesbury Arm was first conceived as a link between the *Grand Junction Canal* and the *Wiltshire and Berkshire*

The seventeen narrow locks on the *Grand Union*'s branch down to Northampton are notorious among canal boatsmen; they prevent craft of more than seven feet wide travelling from the *Grand Union Canal* to the lovely reaches of the River Nene, and thus to the Fens.

Canal at Abingdon on the River Thames—a link that was never built west of Aylesbury. The length which was built (it opened in 1815) remained purely a dead-end branch from the *Grand Junction* down to Aylesbury. As far as the water supply engineers are concerned, the Arm is a dead loss, drawing off valuable water from the main line. Indeed, the canal not long ago was on the verge of becoming unnavigable, but fortunately an official programme of dredging and lock repair saved the Arm just in time. Since then, an ever-growing colony of pleasure boats at Aylesbury Basin has provided sufficient traffic to keep the canal reasonably well-used.

This Arm leaves the *Grand Union* main line at Marsworth, and provides an immediate contrast in its top lock, which is part of a narrow two-step staircase. The canal falls steeply away from the Chiltern escarpment; the close towpath hedge, the very narrow bridges, and the occasional abandoned farm wharf—and even what was obviously once a remote canal pub—all make it every inch the rural narrow canal. About three miles west of Marsworth, the landscape opens out somewhat as Aylesbury comes into view. A couple of locks and an extremely narrow bridge lead to the basin which is lined by pleasure boats. A weir allows the lockage water to escape to the valley of the River Thames.

THE SLOUGH ARM: This was one of the last important lengths of canal built in Britain, only completed in 1883. It is a straight, wide, business-like stretch of water five miles long and of minimal intrinsic interest, there being

73

little of note to see either on or from the canal. An intriguing scheme respecting this canal has been the plan to extend it southwards to the River Thames near Windsor, less than two miles away. This would give the *Grand Union Canal* a useful link with the non-tidal river, but as it would involve negotiating the main railway line to the west and the M4 motorway and Agar's Plough (the playing fields of Eton College), it seems unlikely that this will ever happen.

The Arm leaves the main line at Cowley by an enormous rubbish tip where all the dredgings from London's canals are dumped. A flurry of aqueducts take the canal over a muddle of small rivers and gravel pits (these were the principal reasons for constructing the branch in the first place). There are plenty more signs of gravel extraction further along. A cutting takes the canal past Iver, and then it passes through semi-built up and industrial surroundings before petering out at a terminus in a rather dingy part of Slough. The Slough Arm is not used very much—there is no commercial traffic now—and part of the canal becomes very weedy in summer. However, the boatyard at Iver is useful for pleasure craft, for it is one of the few canal boatyards anywhere near London.

THE PADDINGTON ARM: This important spur, opened in 1801, was the *Grand Junction's* line into London, leaving the main line at Bulls Bridge, Southall, and running level all the way to a large basin at Paddington, at that time the north-western end of the metropolis.

Once this Arm was entirely in the countryside; now it is within west London's outer tentacles, passing through the less salubrious parts of Greenford, Alperton and Willesden. The Arm has always been an important source of canal traffic right up to modern times, for it is not too far from the Thames and London Docks, and its dimensions are quite adequate for barge traffic. So it is not surprising that since its opening in 1801 the Paddington Arm has generated a little belt of industry. During the nineteenth century, private basins and docks began to sprout all along the waterway, and plenty of wharves as well. But these have now fallen into disuse: for example, neither H. J. Heinz nor Ovaltine Ltd any longer use the canal for freight purposes; nor does the power station at Willesden, although it straddles the water. Even the basin at Kensal Green Gasworks no longer receives coal by boat.

This relatively recent flight of trade from the Paddington Arm has left it looking rather depressed and not yet adjusted to its new circumstances. The south end, it is true, has been opened out and smartened up as a pleasant footpath. Bridges have been painted and seats provided, but the great basin at Paddington which has not been used for several years—and which repre-

The Paddington Arm of the Grand Union at Westbourne Park—three very different means of transport side by side.

sents a magnificent opportunity for a large mooring site or boating business on a twenty-eight mile lockfree pound—remains ignored by its owners. Part at least of the basin may yet be filled in for building purposes, which would be a great pity.

Elsewhere, the Arm has some unusual sights, from the M40 motorway viaduct which swings out over the canal near Westbourne Park, to the converse of this, a concrete aqueduct over the North Circular road. There is a wood with a somewhat beleaguered appearance, and a suitably rustic golf course at Horsenden Hill. A long, long wall and railings line the richly overgrown and justly famous cemetery at Kensal Green, but the old gate onto the canal which gave access to waterborne cortèges is now firmly shut. Among more general features of the canal are the large number of 'horse-falls'—steps set into the side of the towpath as an escape route for towing horses that had fallen in the cut. And there are plenty of canalside pubs too,

Little Venice, in the heart of London, used nowadays for narrowboat moorings.

although the Paddington Packet Boat, which recalls the passenger 'flyboat' that once ran between Uxbridge and Paddington, is round the corner on the main line towards Uxbridge.

REGENT'S CANAL: The *Regent's Canal,* which now forms the *Grand Union*'s line from the Paddington Branch through London into the Thames in Dockland, was once an independent waterway. Opened in 1820 at the formidable cost of nearly three quarters of a million pounds (for a waterway only eight and a half miles long), it might never have been a canal, because an original alternative plan suggested a horse tramway instead. But the canal plan won the day and it became an extremely useful extension from the *Grand Junction*'s Paddington Branch down to the London Docks and the River Lee.

The *Regent's Canal* that most Londoners know is the stretch from 'Little Venice'—more accurately known as 'Paddington Stop' in the canal trade— to London Zoo, through to Camden Town and the top of the locks.

Sturdy but elegant canal architecture near Leighton Buzzard, *Grand Union Canal*.

This handsome stretch of waterway embraces not only a genteel, tree-lined length through 'Little Venice' but also the Maida Hill Tunnel and a long green cutting beside the northern edge of Regent's Park. The Zoo is either side of the canal here, with the modern aviary and sundry animal houses right at the water's edge. Trip boats ply up and down this length all through the summer, and since the opening of the towpath to the public from Paddington and beyond to the Zoo and now to Camden Town, it has become a favourite walk for local residents and visitors too. A large colony of noisy and resilient ducks completes the image of a length of municipal parkland.

East of Camden top lock, the *Regent's Canal* becomes less decorative and a little more 'realistic'. The canal's many large trip boats rarely bother to venture down the twelve pairs of barge-sized locks, whose opening hours are rigorously controlled by a team of lock keepers. Indeed, the whole operation of the canal is more geared to the defence of the locks and canal-side premises against vandals and other undesirables than to the free navigation of boats; even the towpath is officially out of bounds. All of this leaves the *Regent's Canal* east of Camden something of a ghost waterway, and all the more enticing for that. The surroundings are a great contrast to those on the top pound. It is very much an enclosed urban canal, a narrow corridor through miles of factory land and office back walls. Here and there the view broadens at the great basins that jut out from the canal. Almost without exception these acres of water space are utterly unused, and most are threatened with infilling for building land. Unfortunately, the timber traffic which has sustained the canal for so long seems on its last legs now, but the occasional barge is still dragged up the canal, either by the lock keepers in their tiny tractors or by small tugs. The latter are necessary if the barges need to penetrate the half mile towpath-free tunnel under Islington.

At the south end of the canal in Hackney is the *Hertford Union Canal*, a short, straight cut dropping down three locks to form a link with the River Lee. The *Regent's Canal* terminates in a great basin which, as the Regent's Canal Dock, used to be a major point of transhipment between ships and canal boats. But the Dock was closed in May 1969, and now serves only as access to the Thames. Only one wharf is still in business, to allow the occasional ship to load here with scrap metal.

PLACES TO SEE THE GRAND UNION CANAL

On a canal as long and diverse as the *Grand Union*, it is surprising how few large towns one encounters, and how painless such encounters are. It is true that the long hauls through the outer regions of London and Birmingham are very disheartening, but the journey through the innermost part of each of these cities is of great interest—and on the top end of the *Regent's Canal* of great beauty as well. In Leicester, the navigation forms part of the 'landscaped' municipal scene, while in the Chiltern towns like Berkhamsted and Hemel Hempstead—and at nearby Leighton Buzzard—the waterway is agreeably open to view. Loughborough, on the Leicester line, is not particularly appealing from the canal, yet the only real disappointment is at Leamington Spa, where the town's sole interest in the navigation which visits it appears to be as a linear rubbish tip.

But the few towns along the way account for only a tiny percentage of the *Grand Union*'s mileage; the rest is countryside, and most of it very green, soft and thoroughly English countryside. There is no difficulty in finding places of canal interest along the way; the problem is to limit the list. A brief, and necessarily rather arbitrary, round-up is given below, starting from the south end.

The *Regent's Canal* is well worth walking, at least along the re-opened towpath between Camden and Paddington (Little Venice). Some bits in Islington are more interesting if less well-known—as is the old *Regent's Canal* Dock in Limehouse. The Paddington Arm is also good for a walk in the Kensal Green area, and the canalside installation opposite the cemetery is a much-vaunted example of vintage 'gasworks architecture'. On the main line of the *Grand Union*, Brentford Depot displays a modern commercial

Brentford Depot, *Grand Union Canal*. There is still a limited amount of commercial traffic on London's canals.

Lockgate workshops at Marsworth, on the *Grand Union Canal*.

use of the canal, despite the fact that the actual amount of transport by canal is minimal (most goods arrive by Thames lighter from London Docks). Heading north, Cassiobury Park outside Watford provides a delightful setting for the waterway. On the Tring Summit, the eighteenth-century canal workshops at Bulbourne—where wooden lock gates are made to this day—are close to the junction with the old Wendover Arm, the Marsworth locks and reservoirs, and the junction with the Aylesbury Arm. There is a handsome group of three locks by the road at Soulbury; and the canal aspect of Cosgrove village is of interest, with a boat-building yard, an unusual Gothic-type bridge, the overgrown entrance to the old *Stratford and Buckingham* Branch, and just to the south the iron trough of Wolverton Aqueduct over the River Great Ouse.

Further north is Stoke Bruerne, with its fascinating and deservedly famous Waterways Museum, and a flight of seven locks and an old double-arched bridge built to cater for the paired broad locks. To the north and west is Braunston whose tunnel, six locks and some typical old canal cottages and boat repair buildings at the foot of the locks merit a visit, quite apart from the boating activities by the reservoirs, or the historical interest in the original course of the *Oxford Canal* hereabouts. At Calcutt near Napton, Stockton, and indeed all the way down to Leamington, there are the 'new', wide locks built in the 1930s, often with the old narrow chambers alongside. But the most impressive place to see these locks is at the Hatton Flight of twenty-one just west of Warwick. The five at Knowle are also worth a look, with their great, square, intervening pounds.

On the Leicester line, Watford and Foxton are obvious places to see the canal, although Foxton is easily the more rewarding. The overgrowth covering the remains of the inclined plane has in recent years been largely cleared, and the old rails can be easily seen. Leicestershire County Council have built a splendid car park and picnic area beyond the top of the locks, and a very pleasant woodland walk leads from there to the canal. Further north, the *Soar Navigation* is more notable for a series of consistently pretty villages than for any engineering works of note.

BOATING ON THE GRAND UNION CANAL

With the exception of part of the Leicester line, all of the *Grand Union* is now wide-locked—which is just as well, since it is a popular mooring among boatowners in the South-East. But this, inevitably, makes it a slower business to get through the locks, especially as many of the top gate paddles have been removed in the past few years. The long climb over the Chilterns is an arduous and frustratingly time-wasting journey for the solo navigator. There are two long tunnels at Braunston and Blisworth, with plenty of room inside for narrow-beam boats to pass; broader-beam craft must give notice of their intention to navigate the tunnels.

The locks are all paired on the *Regent's Canal,* but this has never prevented the authorities from greatly limiting the times that boats can pass through the flight. Fortunately, the duplicate lock of each pair is now being converted to an extra overflow weir which should eventually allow pleasure boats to lock through without being supervised by lock keepers.

A very different flight of locks is to be found at Foxton, where the two staircases of five locks each are separated only by a short pound or passing place. It usually pays to think carefully before embarking on this flight, and it is, of course, essential to check whether there are already any approaching boats in the staircase.

The channel on the Leicester Summit is often narrow and winding, so a good lookout for oncoming boats is advisable. Happily, however, much dredging in recent years has trimmed back the great banks of weed that used to take up much of the channel.

Foxton Flight on the Leicester Arm of the *Grand Union Canal*—views up and down the flight, which consists of a double staircase of five locks, a waiting pound, and then five more. (See the remains of the inclined plane to the right of the view down.)

4

East Midland Waterways

RIVER TRENT

EREWASH CANAL

NOTTINGHAM CANAL

GRANTHAM CANAL

FOSSDYKE AND WITHAM NAVIGATIONS

CHESTERFIELD CANAL

The Trent may not be the most handsome river in England, yet its broad and sheltered reaches sweeping northwards from Nottinghamshire down to the Humber have not only provided for centuries a water highway from the East Midlands to the sea, but have over the years fostered the growth of many artificial canals. It is thanks to these canals that the Nottinghamshire–Derbyshire coalfield was opened up so thoroughly. To the east of the tidal Trent is the ancient *Fossdyke Canal* providing a link with the River Witham and the Wash.

River Trent

The Trent is the river that drains most of the East Midlands. It is also a navigation of generous proportions and the primary waterway route for the whole region into which seven other navigations flow. The river's large scale accentuates the difference between it and any artificial waterway, a difference which is not so obvious on, say, its navigable tributary the Soar.

The river's character changes as it flows north, but there are certain features which seem to characterize the Trent wherever it is found. Among these are the power stations which have sprung up at intervals along the waterway almost from end to end. Their belching chimneys, and the giant cobwebs of transmission cables emanating from them, are quite out of keeping with the soft green valley that typifies this particular river; for the Trent Valley is by no means spoilt, at least in the sense of being built up to any appreciable degree. It is primarily a flat, green pastureland dotted with small farming villages. Here and there are hills, woods, and even towns, but mostly the Trent is quiet and undramatic, winding this way and that through the landscape. It has a name for being heavily polluted—at the locks the adjacent weir is often hidden by a wall of detergent foam—but this does not stop the energetic from sailing, rowing and even waterskiing up and down the river. Nor does it deter the inhabitants of the surrounding area from descending upon their favourite spot to enjoy a picnic or paddle in the water. Its open, grassy banks and soft, sandy shores are well suited to this.

The river's source is right up in the hills north of Stoke-on-Trent. From there it heads south through Staffordshire, curving round in a great horseshoe through southern Derbyshire and Nottinghamshire before heading north-wards at Newark and straightening out through Lincolnshire to join the Yorkshire Ouse at Trent Falls. The two rivers combine to form the Humber Estuary. Just below Shardlow the river is canalized in several places, there being long-ish lock cuts for both Sawley and Cranfleet locks. At Beeston, the navigation quits the river altogether, being directed down the short and rather scruffy *Beeston Canal* (or Cut) to join the *Nottingham Canal* at Lenton and eventually ending up back in the river near Trent Bridge. But from here on, the river and the navigation are the same entity all the way down to the Ouse, except at Newark where the river channel divides, one line going over a weir to the west and the other making for the heart of the town. Having temporarily lost much of its width, the navigation fits comfortably into Newark, and the two channels join up again at the north side of the town.

At Shardlow, near the start of the *Trent and Mersey Canal.*

The River Trent at Gainsborough.

Five miles north of Newark is Cromwell Lock and the start of the tideway. The river continues as before—if anything, a lot more winding in its course—through a landscape almost empty of features except for the very occasional large bridge or power station. Downstream of Nottingham, the river has been kept well up to date during the present century, with several of the locks greatly enlarged, mechanized and equipped with overhead electric traffic lights. But the Trent is sadly underused by commercial traffic, despite the substantial size of craft it can take. Petrol tank barges used to ply up and down with oil from Immingham bound for the oil storage depot at Colwick (Nottingham). Nowadays, the oil is carried in railway wagons. Even general merchandise barges are becoming scarce, and not one of the power stations receives its fuel by water. It is only on the tidal section that much activity takes place; small coasters carry grain up to Gainsborough, and bigger ships still come to Gunness Wharf bringing iron ore for Scunthorpe. But the now-fixed bridge at Keadby allows only sixteen feet of headroom at high water.

Bridges are few and far between on the tidal river. The settlements on each bank were once connected by small passenger ferries, but since the age of the motor car these have fallen into disuse and now there is virtually no communication between either side of the river, except in the vicinity of the bridges. However, there is often a minor road running along one side or the other connecting the various hamlets that punctuate the banks, and although it is broad and isolated, the lower Trent is never a lonely waterway.

BOATING ON THE RIVER TRENT

The upper Trent is a popular boating water, especially for day or weekend trips—its wide course, slow current and small number of locks make it ideal for this (although canal boaters will probably find it rather dull, and lacking in places to moor up). The Trent is also extremely well suited for rowing and sailing activities. Sawley and Cranfleet are very busy boating areas, with hundreds of craft based at either boat clubs or boating firms. The *Erewash Canal* and the River Soar are both immediately accessible from here, which makes it an excellent area for mooring.

Beeston (at the west end of the Cut) is also a well-frequented mooring site, and so is the far end of the *Nottingham Canal* at Trent Bridge. Each river lock, of course, is accompanied by a weir, which often lurks unseen at the back of the lock island. The weirs are unfenced, but the lock cuts are well signposted. Most of the lock chambers below Nottingham are not only very long (165 feet and more) but also very deep. However, the fall is not very great, and it is not always necessary to tie a boat up when ascending the lock. The keepers are usually well aware of the apprehension of boaters going up one of these big locks and, consequently, they are inclined to go gently with the powerful hydraulic paddles.

The tidal Trent is a very different proposition. It is best used only as a link between one navigation and another, for there is hardly anywhere to moor up even for a very limited period. It is essential to be fully informed as to the tide times.

Erewash Canal

This is a short canal of eleven and three quarter miles, extending north from the river Trent up the shallow valley of the river Erewash. Although a minor river by any reckoning, the Erewash was an obvious source of water power for local industry long before the canal was first projected in the second half of the eighteenth century. The prime reason for building a navigation along the valley at this time was to serve the various collieries exploiting the rich coal seams that were found at the north end of the valley, around Eastwood, Ilkeston and Langley Mill, in the second half of the eighteenth century.

Not surprisingly, it was mainly coal owners who initiated and financed the canal, and they were energetic about its construction. Once the project had received Parliamentary approval in 1777, the whole work was completed in a little over two years. The canal had the desired effect, enabling Derbyshire coal to be shifted cheaply and efficiently to markets down the Trent and up into Leicestershire, and even more profit began to accrue when, in 1792, the *Cromford Canal* extended the *Erewash* line northwards by another fourteen and a half miles. Another lesser extension was represented by the *Nutbrook Canal,* which was built later as a spur leading north and west from the *Erewash* at the great Stanton Ironworks. The canal had its competitors—the *Nottingham Canal* which followed the very same valley, and later the Midland Railway, which originated at Eastwood. But the *Erewash* stuck it out until it was absorbed by the Grand Union Canal Company in 1931. Technically, it is still part of that line but, as with the *Soar Navigation,* its separate identity is usually acknowledged.

Being a canal that follows a river valley, rather than a contour canal making its own way through a landscape, the *Erewash Canal* treads a relatively predictable path, with a series of locks lifting it north from the Trent. The canal's surroundings could be described as 'light mining industrial' and show plenty of signs of its coal-mining background although the canal is never overwhelmed by these excesses or dominated by slag heaps. Likewise, the concentrations of industry in the Long Eaton area are never so intensive as to cast a cloud over the waterway, for this is a canal which always seems to be changing its character and surroundings, dodging from one place to the next, passing one minute through a housing estate, and the next through watermeadows. Other identifying characteristics include the subsidence that has conspired to lower many of the bridges on the northern section, and—more happily—the many delightful canalside gardens that are so prominent along this route.

The canal locks up out of the River Trent at a waterway 'crossroads' opposite the mouth of the River Soar. Away from the junction, it runs beside a tree-lined road into Long Eaton. At the other side of Long Eaton lock, the River Erewash—a dirty, nondescript stream—arrives to dog the canal for the rest of its journey north. The enormous expanse of Toton railway sidings—the main marshalling point for southbound coal from the local collieries—introduces the Midland railway to the canal and they share the valley from this point northwards. The passage through Sandiacre is enlivened by the presence beside the canal of some large, handsome textile

Trent Lock, the southern terminus of the *Erewash Canal* where it meets the Trent River. This is a very popular spot in summer, not just for boating—there are pubs, shops, car parks and a playground.

mills. At Sandiacre Lock, the remains of the *Derby Canal* used to join the *Erewash*; but the former is dead now and the junction lock is abandoned. The belching chimneys north of Sandiacre represent the great roaring furnaces of Stanton Ironworks. Quite unidentifiable now is the one-time junction with the *Nutbrook Canal,* which used to run right through the works and whose now-piped water still cools the plant here. Just across the valley is the old *Nottingham Canal* snaking along the contours. These two canals stay very close, sharing the same valley. At Ilkeston, the *Erewash* goes under a great iron trestle railway aqueduct before emerging into rather more open countryside, and from here up the canal wears a rather carefree, rural air. Horses graze quietly in the fields, an old narrowboat butty decays gracefully to one side of the waterway, and water birds rustle around in the canal banks. Even the well-worn, crudely cut balance beams at the locks reflect the mood.

But all this comes to an end at Langley Mill with industry, a shabby

89

Great Northern Basin, Langley Mill, the northern terminus of the *Erewash Canal*. At one time this was the junction of the *Erewash* with the *Cromford* and *Nottingham* canals, but the latter two are now filled in at this point.

townscape and a large, noisy new road. The canal rises through its four-teenth and final lock into the Great Northern Basin, an odd triangular basin which has recently been completely restored, as has the lock. Its present condition is entirely due to the efforts of the Erewash Canal Preservation and Development Association, which in recent years has done a great deal to revive and restore the waterway. The basin is technically part of the *Nottingham Canal,* but around Langley Mill other traces of both the *Nottingham* and *Cromford* canals, which once joined the *Erewash* at the Great Northern Basin, have been virtually obliterated.

Nottingham Canal

Nottingham, as well as being situated on the Trent, has long been the focus for various minor canals connected to the river and to each other. Two of these—the *Beeston Cut* and the *Nottingham Canal*—are of necessity part of the through-route of the *Trent Navigation*. This unexpected diversion for several miles from the course of the river itself is caused by a long-standing shoal in the river bed, sometimes known as the Clifton Fault. There have been suggestions from time to time of blasting this shoal and restoring navigation to the river channel, but in the meantime, boats will have to use the *Nottingham* and *Beeston* canals to get around the obstacle. It is a drab and somewhat depressing diversion.

The *Nottingham Canal* was opened in 1796, connecting the river Trent (at Trent Bridge, Nottingham) to Langley Mill, the junction with both the *Erewash* and *Cromford* canals. Designed principally as a supply route for coal into the city, and once a strong rival of the *Erewash,* the *Nottingham Canal* today is very much a lost cause. Although much of its long northern pound is still in water and makes, indeed, a pretty and winding waterway as it follows the contours of the Erewash valley (the two canals share the valley for several miles), its course near Langley Mill has been obliterated— as indeed it has through most of Nottingham. The flight of seventeen locks that once lifted the canal from Lenton to Wollaton has now all but vanished under the spreading suburban acres of Nottingham's new housing. So, of the canal's original length of almost fifteen miles, only two and a half remain navigable—from Lenton to Trent Bridge. Various of the canal's short branches have also disappeared, except for the old Poplar Arm, near Nottingham Station, which lies disused but not yet abandoned. Like most of the surviving length of the *Nottingham Canal,* the Poplar Arm is hemmed in by heavily built-up surroundings.

The *Beeston Cut* runs from the *Nottingham Canal* at Lenton to the River Trent in Beeston. It is nondescript and runs through the bleak industrial hinterland. There is one lock into the river above the weir at Beeston, and another (disused) at right angles which used to bypass it, allowing boats into the reach (which is not now regarded as navigable).

Grantham Canal

The *Grantham Canal* is a delightful old contour canal meandering eastwards from the Trent at Nottingham towards Grantham. Once out of West Bridgford, the canal wanders off through a string of sleepy villages on the fringes of Nottinghamshire, Leicestershire and Lincolnshire, contributing to the quiet and unassuming beauty of the Vale of Belvoir. The great Belvoir Castle looks out from a long ridge towards the south where comfortable, settled, unspoilt countryside, laced together by a network of minor country lanes, seems to have escaped the worst side-effects of the Motor Age. The single blot on this landscape is Cotgrave Colliery which was built beside the canal about a decade ago.

A possible candidate for restoration—one of the Fosse Locks on the *Grantham Canal, 1974*.

The *Grantham Canal* is not navigable: all eighteen locks are derelict, and most of the bridges have now been culverted. The canal, however, still holds a full level of (clean) water for most of its length and is a good example of a rural canal which it would be pointless and expensive to eliminate, although it serves little useful purpose in its present state. It is likely that before long the canal will be 'tidied up' and fitted for some suitable recreational role; it is even possible that in the very long run it might be restored almost to Grantham. Supporters have drafted a scheme to cut a completely new connection with the river Trent at the canal's western end to avoid the obstacles that would be encountered at West Bridgford. This revolutionary but practicable solution is probably the only potentially successful chance of restoring this waterway as a through route. Meantime, the canal sleeps on.

Fossdyke Canal and the River Witham

These two waterways are normally treated together because they form a single line of navigation. Between them they connect the tidal River Trent at Torksey with an inlet from the Wash at Boston, traversing an almost entirely dead-flat stretch of sparsely populated but intensively cultivated fenland.

Both waterways are physically similar, but historically very much in a separate class from most inland navigations, for both are very long established waterways. The *Fossdyke* (or Foss Dyke) is believed to have been built in Roman times, and the Witham is generally assumed to have been a navigable (and tidal) river since well before the Norman Conquest. (The *Car Dyke,* which runs parallel to the Witham for many miles starting just east of Lincoln, is also known to have been cut by the Romans.) Both navigations doubtless contributed much to Lincoln's prosperity in the Middle Ages although both have had a very chequered history since that time.

Today, despite the generous dimensions of these waterways and their proximity to the port of Boston there is no longer any commercial traffic along the route. But full use is made of it by boaters, both for cruising and for sailing. The Witham is also extremely popular among anglers, and its banks tend to draw many campers and caravanners. In winter, however, it is a very different story: the Witham is an important drain for a large area of Fenland, and the various dykes that flow into it can, in wet weather, raise the level of the Witham considerably. In such conditions, boats may not moor on the river. The Witham's drainage function explains not only all

93

Near Boston, on the *Witham Navigation*.

the tempting dykes that lead off the river's course (these have one-way gates that close automatically if the level of the Witham rises much above that of the branch waterways) but also the high grass banks that flank the river's wide, straight course. These banks hem the river in virtually all the way from Bardney to Boston and they do nothing to improve the view from the river itself—although the only view is of miles of the flattest fenland; fertile, intensively cultivated, and very short of landmarks. The occasional riverside pub, a handful of bridges and a number of anglers and swans are all one is likely to meet.

The *Fossdyke* lives up to its Roman origins all too well. A succession of wide, straight reaches links the River Trent with the Witham in Lincoln. The banks are somewhat lower than on the Witham, but there is still little of interest, save for the village of Saxilby. There are few bridges and only the tidal lock at Torksey.

It cannot be pretended that the *Fossdyke and Witham Navigations* are particularly fascinating waters. In approximately fifty miles there is precious little to see from the straight wide waterway. However, the passage through

The famous 'Glory Hole' in Lincoln. This waterway forms the west end of the *Witham Naviga-tion*.

Lincoln is a great contrast to the flat open waterways on either side. The *Fossdyke* meets the Witham at Brayford Pool, a great stretch of water near the centre of the city. (The Witham flows in from the south.) Just to the east of the junction is a very narrow and fast-flowing stretch of river which is straddled by the famous High Bridge, a medieval structure stacked with old black-and-white timbered houses and usually known as the 'Glory Hole'. Further along are some warehouses which were supplied by water until recent years.

For many people, the journey through Lincoln is the principal reason for navigating the waters of the *Fossdyke and Witham*. For others it is to visit the handsome and equally historic town of Boston. A handful of boatmen use the route as a short cut between the Wash and the River Trent; and some go to unravel the geographical mysteries of the *Witham Navigable Drains*—a complicated network of waterways stretching to the north of Boston. The *Drains* are well-named, for their navigability is guaranteed by no one and the water level depends entirely on the weather. In too dry weather there is a risk of insufficient water, and when the drains fill up in wet weather the headroom under bridges shrinks to almost nothing. The central junction of these waterways is at Cowbridge, a mile or so north of Boston. The dykes fan out northwards from here for a distance of up to ten miles and as far as the south end of the Lincolnshire Wolds, along waterways known by such names as 'Stonebridge Drain', 'Hobhole Drain' and 'West Fen Catch-water Drain', with the 'Maud Foster Drain' running down to Boston. They add up to a fascinating microcosm of the Fenland waterways further south nearer the rivers Nene and Great Ouse, but the *Witham Drains* are even less well known than the Fens, and it is rare to see a boat upon them.

One other formerly navigable canal leading off the River Witham was the *Horncastle Canal,* a waterway built purely for the purpose of navigating from the river near Tattershall Castle to the town of Horncastle itself; but this has now been abandoned and its entrance blocked off.

BOATING THE FOSSDYKE CANAL AND THE RIVER WITHAM

A wide, straight waterway passing through dead flat and mostly featureless countryside, the *Fossdyke and Witham* route does not appeal to all boaters, although a visit to Lincoln or Boston makes the trip worth while, and it is easy to make good time along the waterway. Apart from the tidal locks at either end of the canal, at Torksey and Boston, there are only two other locks in over forty miles; namely Stamp End Lock (in Lincoln) and Bardney. Both are operated by lock keepers. The main line is suitable for fairly large craft, but those who wish to explore the *Witham Navigable Drains* would do

Tattershall Castle rises a short distance from the *Witham*.

well to take a reasonably short and narrow boat of low profile. The entrance lock to the system is at Antons Gowt, two and a half miles north-west of Boston Grand Sluice; the only other useable lock within the network is at Cowbridge.

A point to watch is Lincoln. Brayford Pool is very wide, but rather shallow away from the dredged through-channel that runs along the north side.

Brayford Pool in Lincoln, urban water space whose potential has not been fully exploited.

Chesterfield Canal

In some ways the *Chesterfield Canal* bears some affinity to the *Ashby Canal*. As on the *Ashby,* the coal mining areas towards the canal's terminus, once so important to the canal's early prosperity, proved to be too much of a good thing; and also like the *Ashby,* the subsidence which led to closure of the top end left a canal of a purely rural character ideally suited for the pleasure boating age.

This canal is something of a hybrid. Although, like most canals leading directly off a river navigation, it was built wide enough to take at least the smaller of the regular traffic using the river navigation, the locks are wide for only part of the distance, and in any case the channel alone is something less than generous in dimensions. But although this cannot have helped the boatmen of old when two fifty-ton boats met on a corner, it has certainly helped to invest the canal with an intimate character.

The *Chesterfield Canal* was built fairly early during the canal age: James Brindley was one of its principal instigators and its first engineer. But he died

One of the narrow locks on the little-known *Chesterfield Canal*.

in 1772, only a year after construction had begun, and the work was finished in 1776 by Hugh Henshall, Brindley's brother-in-law. The prime purpose of the canal was to provide an outlet for the coal and iron produced in areas of the Rother valley just to the north and east of Chesterfield. Before the canal was cut, most of this traffic for distant destinations had been carried by pack horses to Bawtry and loaded there onto boats on the River Idle which joins the River Trent at West Stockwith. But it was a long slog by road, and an uncertain ride by boat. A proper man-made canal was clearly preferable, enabling goods to be loaded directly into narrowboats for onward shipment up the Trent and its tributaries into the south Midlands and via the 'Grand Cross' of waterways to the other three principal river estuaries of England. Hence the *Chesterfield Canal,* and hence its commercial success.

It was not a particularly expensive canal to build—it cost about £152,000 in spite of containing sixty-five locks from end to end and a tunnel over 3000 yards long. Today the west end of the canal—the twenty miles between Chesterfield and Worksop—is in a sorry state. The collapse of the great tunnel at Norwood in 1908 signalled the end for the top fourteen miles and severed Chesterfield's waterway link for ever. Subsidence from the collieries once served by the canal also took a heavy toll of the forty-nine narrow locks that once lifted the waterway over the hills, down to the Rother valley and the massive Staveley ironworks, and then up to Chesterfield. Today the locks lie abandoned and forgotten. It is no coincidence that the remaining twenty-five miles of the canal, from Worksop down to the Trent, are only lightly locked although they pass through no mining area. They are entirely navigable, although it is worthy of note that until recently, even this most attractive section was decaying and faced with possible closure: it is now assured of a certain future.

THE COURSE OF THE CANAL

The surviving section of the canal winds its way through the rather under-populated farmland of the northern edges of Nottinghamshire, passing several quiet villages and towns. The start of the canal at the junction with the Trent is misleading. A large lock gives access up from the river into a basin which provides sanctuary for plenty of river-based boats of various sizes. The first stretch of the canal proper is straight for half a mile, to the village of Misterton where the first of several wide (fifteen feet) locks are to be found. From here the canal begins to wind through the quiet countryside, following the contours of the slight ridge on the south side. Then it turns sharply south

An old Pickford's warehouse over the *Chesterfield Canal* in Worksop.

to cut through a short if unexpected tunnel carved out of the rock. Running through a short stretch of parkland and around two villages, the canal follows the barely-defined valley of the River Idle as far as Retford, where it crosses the river before making for another minor river valley, this time the Ryton. From Retford onwards the locks are narrow—obviously the river craft would have got no further than this. The narrow locks were favoured by the canal's builders not only for their relatively low, original construction cost, but also for their economy in the use of water. West of the A1 dual carriageway road, the canal passes yet another stretch of private parkland at Osberton before crossing the River Ryton on a tiny aqueduct and entering Worksop, where there is at last a whiff of the coal mining industry. The navigation is now limited by a derelict lock just west of the town centre.

PLACES TO SEE THE CANAL

The early date of this contour canal and the relatively easy terrain between West Stockwith and Worksop precluded the need for engineering works on any kind of major scale for the navigable part of this canal. (The un-navigable section features the now-subsided Norwood Tunnel.) The aqueducts are insignificant and the cuttings and embankments minimal. The bridges are an interesting mixed bag whose wide range of styles speaks of considerable latitude being exercised during construction of the waterway. So the interest of this canal is more in the surroundings, the wildlife and occasional details like the setting of some of the lock cottages. More specific places include the terminal basin at West Stockwith, the strange little tunnel at Drakeholes, with the handsome pub at the south end, the environs of Osberton Lock and Park, and the rare and intriguing example of an old warehouse straddling the canal in the centre of Worksop.

BOATING ON THE CANAL

The *Chesterfield Canal* boasts less than its due of cruising boats because, although one end of it is quite close to the heart of England, the other end abuts the tideway of the River Trent, a waterway which many boat owners used to narrow canals are hestitant to navigate. This isolates the canal from the rest of the network; but the waterway is in fact ideal boating water for smaller canal boats. The water can be a little shallow in places, and there is a shortage of winding holes, so it is better to have a boat not more than about forty feet long. The locks are wide (about fifteen feet) up to Retford, and narrow thereafter. All are unmanned, except the one at West Stockwith which connects with the tidal Trent. Based at Clayworth is the Retford and Worksop Boat Club, a helpful and hospitable place which has been active in looking after and restoring the canal. The canal's main boating business is at Retford, but there are currently no hire-cruiser firms of substance operating on the canal, probably because of its limited length.

5

Yorkshire Waterways

RIVER OUSE

RIVER DERWENT

POCKLINGTON CANAL

AIRE AND CALDER NAVIGATION

RIVER AIRE

SELBY CANAL

CALDER AND HEBBLE NAVIGATION

HUDDERSFIELD CANALS

SHEFFIELD AND SOUTH YORKSHIRE NAVIGATION

NEW JUNCTION CANAL

The broad, flat basin of southern Yorkshire is veined by a large number of substantial rivers, many of which have been navigable on the tide since time immemorial—the Ouse, the Derwent, the Don, the Aire and the Trent. Their co-incidence in this region, near to the sheltered coastal ports, has made it a thriving, self-contained trading area for centuries. Man's pushing back the head of the navigation up towards the hills has created an extensive system of canals and canalized rivers, and these now form virtually the only group of inland waterways carrying any quantity of goods in this country.

TRAFFIC ON THE YORKSHIRE WATERWAYS: The waterways of southern Yorkshire, close to big open tideways and well placed to benefit from economies of scale, still function as important transport routes. The *Aire and Calder* route, the backbone of the system, is heavily used by tank barges which fill up with fuel at Immingham on the Humber and take it as far up as Ferrybridge, and further still to Leeds and Wakefield. It also carries coal from the collieries to the power stations along the navigation, although even in this region of giant power stations, this traffic does not add up to

104

'Tom Pudding' coal hoppers waiting to be unloaded into a ship's hold at Goole Docks, where the *Aire and Calder Navigation* meets the River Ouse.

full capacity; general merchandise makes up the rest of the cargoes. Other inland waterways still carrying freight are the *Sheffield and South Yorkshire* and part of the *Calder and Hebble,* although the traffic on both of these is decreasing. They have suffered from lack of modernization and their commercial future is uncertain. Even the non-tidal reaches of the principal river in the area, the Yorkshire Ouse, are very quiet commercially.

The most important form of 'boat' operating on the canalized rivers of the region has long been the 'Tom Pudding', an ancient system of containerization. Patented back in 1862 by William Bartholomew of the *Aire and Calder Navigation,* it proved to be an immense success, and the system still operates virtually unmodified on the *Aire and Calder* and the *Sheffield and South Yorkshire.* It comprises a string of square floating boxes (known as 'Tom Puddings', compartment boxes or pans), capable of holding up to forty tons of coal each. The boxes are lashed together and towed (they were originally pushed) by a tug in a long 'train'. A false bow is attached to the box behind the tug to streamline the train which, upon arrival at its destination, is discharged by a large gantry which heaves each box in turn out of the water and inverts it into a hopper, or even directly into a ship's hold, as at Goole docks. This system offered obvious economies in manpower, time and capital invested, and it has never been replaced as the mainstay of all South Yorkshire's waterborne coal traffic. The compartment boats, or pans, can be seen anywhere on the lower reaches of the *Sheffield and South Yorkshire* or the *Aire and Calder.*

In recent years, a new version of the compartment boat has been designed—a rectangular steel craft carrying not just forty tons but over 160 tons. Three of these are fixed together and propelled by an outboard-engined tug. This is the system that serves Ferrybridge Power Station, but although its labour costs are very low, the capital cost of providing a hoist with a hopper and crane on such a large scale is very high—a factor which tends to inhibit the spread of the system.

Other traffic regularly found on the South Yorkshire waterways includes an ever-shrinking fleet of motorized barges based on the Yorkshire keel sailing barge. These motor barges have long been used for coal and general merchandise: simple economic factors are squeezing them out. To a certain extent they are being replaced by the BACAT (Barge aboard Catamaran) craft that have brought the concept of international containers to waterway transport. The principle with these is to operate a fleet of standardized 'dumb' (i.e. without a motor) rectangular barges which are loaded, filled with merchandise, loaded onto a special 'mother-ship' and carried across the sea—say to Holland—where they are off-loaded at the port and shifted inland by a tug. This last part of the operation is run on Bartholomew's principle

Modern coal transport on the *Aire and Calder*.

of lashing several together at once and pushing or towing them with a single tug. The BACAT system has already begun to operate on a limited scale to as far up as Rotherham, and if the enterprising attitude shown by the British Waterways Board is recognized and encouraged by government planners, the future of these North Eastern waterways—ideal BACAT territory, with medium-sized ports serving medium-sized river-based inland navigations—could indeed look bright. Up to now, various governments have been reluctant to recognize—witness the *Sheffield and South Yorkshire* improvement scheme—any scope for increased use of Britain's commercial waterways, with or without new techniques of handling.

River Ouse

The Ouse is the 'recipient' river of the Aire, Wharfe, Swale, Derwent and Nidd rivers, and it is the Ouse which joins with the Trent to form the Humber, and it was the Ouse up which the Romans sailed to reach York and beyond. A trade route for hundreds of years, its only importance now as a commercial waterway is from Selby downwards, although there still exists a small trade up to York.

The first navigation works on the Ouse were authorized in 1462 when the city of York was granted responsibility for the river. Later Acts of Parliament allowed for the river navigation to be extended first to Linton on Ouse, and then to Swale Nab and on up the Swale, while the Ure—the river whose junction with the Swale formed the Ouse—was made navigable, under an Act of 1767, up to Boroughbridge and Ripon, the last two miles being a canal cut from Oxclose Lock upwards. This effected a continuous line of navigation for seventy-one miles from Ripon to the Humber. Today, the last mile of the *Ripon Canal* up to Ripon itself is disused and the top two locks dismantled, although there is a chance that this length may yet be restored.

Thus, the upper limit of navigation now is near the small village of Littlethorpe, just south of Ripon. The canal drops into the river at Oxclose Lock, and follows it past Newby Park and down through a wooded reach (and a second lock) to Boroughbridge. A canal cut leads to Milby Lock, and the river then winds along down to Swale Nab, the historical start of the River Ouse. From here downwards, the river forms the basis of the 'Vale of York', and since it is a vital drainage course, it is enclosed by high flood banks for much of its way. These have a corridor effect, enclosing the river and shutting it off from the countryside it traverses. This is a pity for anyone on a boat, for the Vale of York is soft, green countryside, quite unspoilt as yet by industrial development or population expansion. There are plenty of villages along the winding river, but although the Ouse is not particularly wide, bridges are few and far between. There is only one between York and Boroughbridge (twenty miles) and only one between York and Selby (eighteen miles). The river has thus always been a natural dividing line through the Vale of York—a division long recognized by being taken as the official boundary between Yorkshire's East and West Ridings.

In York itself, the Ouse courses straight through the middle of the city. Trading vessels still reach this far, bringing grain and other goods up from the Humber ports, but this traffic is trifling compared with the days when the Ouse was the very keystone of York's prosperity. Nowadays, the barges are heavily outnumbered by the rowing eights, fours and endless other pleasure boats both large and small that make the most of the river's delightful open course from York to Naburn.

Naburn Locks, six miles downstream of York, signal the start of the tideway. The character of the navigation changes again, becoming a swift-moving but twisting current that takes thirteen and a half miles to cover the seven and a half miles to Selby. There is a wooden swing bridge at Cawood and two more swing bridges across the river at Selby itself. Selby also contains plenty of coastal shipping that brings grain up to the towering mills that rise on each bank. From Selby the river runs down to Goole, picking up extra width from the waters of the rivers Derwent and Aire. By the time it has reached the new M62 bridge just above Goole, the Ouse is a broad

A tug seen at Selby. These craft are designed more to push and pull from the bow than from the stern.

river sweeping through the flat landscape and carrying all manner of ships and barges. The channel is buoyed all the way down to Trent Falls, at which desolate spot the Ouse and Trent merge to form the Humber, a busy estuary over a mile wide. Hull is just eighteen miles downstream, with Spurn Head and the North Sea around the corner.

SEEING THE RIVER OUSE

The Ouse, with its extension the Ure, is a waterway more for using than for looking at. Being a river navigation, it stays quite separate from the places it passes, and the high flood banks accentuate this effect. There are few locks to impede its course, and the Ouse has none of the intimacy associated with smaller navigations. However, the upper reaches of the River Ure are considerably more open, and stretches such as the reach at Newby Park are positively delightful, with the Queen Anne mansion standing right beside the river. (Bishopthorpe Palace at Naburn is another outstanding building on the river's edge.) The *Ripon Canal* is the northern limit of the inland waterways network and is also a pleasant stretch, but the abandonment of its northern end now denies access for boats into Ripon.

The Ouse is somewhat remote for most inland waterway explorers, so it tends to be used more for local sailing and cruising, with only occasional long-distance pleasure cruisers penetrating from the south. The York–Naburn reach is very busy with sailing clubs, motor boats and plenty of boating facilities at Naburn itself. Upstream, the Ouse is much less crowded, although there are motor boat clubs at Linton Lock and on the *Ripon Canal*. Below Naburn, boats are bound by the tides and there are few places at which it is possible to moor. It is of course necessary, especially on a falling tide, to keep to the outside of the sharp bends and to maintain a good lookout for oncoming craft. Below Goole, the Ouse is wide and busy, and should be treated with particular respect.

River Derwent

Yorkshire's Derwent is slowly coming to life again after a long, dormant period. Its position at the north-east corner of the connected waterways network has ensured that it has never been well known or of any abiding interest to the cruising fraternity, or indeed to many other waterways enthusiasts. But the recent revival of its subsidiary canal, the *Pocklington*, has focused attention on the sad state of the Derwent, and a Trust is now working on a scheme to restore navigation to the river.

The river rises not far from Scarborough and by a fluke of geology turns west instead of east into the sea, running through the low-lying vale of Pickering, between the heights of the North Yorkshire Moors on one side and the Yorkshire Wolds on the other. At Malton, the hills close in on either side, and for several miles the river cuts through a narrow twisting gap between the wooded hills. As the hills gradually recede, the Derwent emerges into the fertile lowlands of the Vale of York and joins the tidal River Ouse near the small farming village of Barmby on the Marsh.

The river, first made navigable to Malton by an Act of 1701, has for many years been navigable only on the tidal section up to the lock between Elvington and Sutton upon Derwent (this lock and the four others—at Stamford Bridge, Buttercrambe, Howsham and Kirkham—having fallen into disuse). But now, following the championing of the river by the Yorkshire Derwent Trust, Elvington Lock has been reopened and the head of navigation moved up to Stamford Bridge. Meanwhile, at the river's junction with the Ouse, an interesting separate development has resulted in the improvement of navigation. The *Yorkshire Water Authority* has constructed a barrage at the mouth of the river—a mouth which has, oddly, always pointed 'upstream' into the Ouse. The barrage is purely in aid of improving abstraction for water supply purposes, but its effect of excluding the tide from the Derwent and the provision of a navigation lock—usable at all states of tide—makes the lower Derwent not only easier to enter but in every way a more relaxing waterway to navigate.

Further upstream, the plan to restore navigation to Malton will, if successful, bring the river's most attractive reaches back into the fold. The main problem is with the locks, all of which are derelict (the one at Kirkham is completely obliterated, having been replaced in 1958 by a modern power-operated sluice). Clearly, an entirely new lock would have to be built; but if this and similar problems can be overcome, and the river dredged where necessary, restoration of the Derwent is likely. It would certainly be worth it, if only to be able to go up this valley by boat to the ruins of Kirkham Abbey.

Pocklington Canal

The *Pocklington Canal* is a short off-shoot of the River Derwent in Yorkshire. A quiet backwater of the canal system, the *Pocklington* is only now awakening from years of neglect and oblivion. Indeed, the fact of its current restoration makes the canal as well known (if only in name) among the canal fraternity as any more central or more useful waterway.

It is unlikely that even in its trading days the canal would have been well known outside Yorkshire, for the *Pocklington* was always a purely local canal. It was promoted as such, without any underlying or grandiose plan to make it part of any important through-route; and after its opening in 1818 it fulfilled an unspectacular role as a means of shipping out agricultural produce from this fertile area in exchange for coal and manufactured goods from other parts of Yorkshire. It was useful to the Pocklington district too, but it was clearly not destined for a dynamic or particularly profitable life, and the canal was sold to railway interests only thirty years after its opening. Not surprisingly, trade dwindled to a mere trickle during the second half of the century, and by the 1930s the canal was unnavigable and had begun to decay. However, the resurgence of local interest in the 1960s led to the formation of the Pocklington Canal Amenity Society, and it is due to the Society's campaign for restoration, and their strenuous physical efforts to clear up the old canal, that the *Pocklington* today is looking forward to a healthier future.

East Cottingwith Lock on the *Pocklington Canal* where it leaves the River Derwent.

The canal is just nine and a half miles long, from East Cottingwith by the River Derwent to Canal Head, the terminus of the waterway on the York to Hull main road (A1079). Rather mysteriously, the canal was never designed to reach the town of Pocklington itself. Near the junction with the River Derwent is Cottingwith Lock whose reopening in 1972 once again allowed boats to reach the canal from the Derwent; from here, the canal runs north and east along the little beck. The bridges are a mixture of low iron swing bridges and some remarkable hump-backed brick structures that incorporate a grand scroll of buttress at each corner. Past the little farming community of Melbourne the canal makes for Bielby, where a short branch goes off towards the village and the main line of the canal turns north for Pocklington, negotiating five of the nine locks in the last two miles or less before the canal's terminus. This most fortunate distribution of the locks enables a maximum length of the canal to be restored for a minimum outlay on lock rebuilding.

A substantially built bridge on the *Pocklington Canal*.

The *Pocklington Canal* is a beautifully quiet rural waterway, tucked away in the north-east corner of the canal system. It boasts no spectacular things to see, and is of little importance to a boater; it is the canal's flavour that gives it its appeal. It is well-endowed with scope for fishermen, walkers and, by no means least, for bird watchers. Wheldrake Ings, on the Derwent at Cottingwith, is an excellent spot, particularly for water birds in winter when the meadows are often completely flooded for weeks on end.

Aire and Calder Navigation

The *Aire and Calder* is the trunk route of South Yorkshire's inland water-ways, and it remains a shining example of how at least one of Britain's early river navigations has managed to keep up with changing technologies, markets and scales. It even makes a profit, and still more encouraging is the fact that the waterway is constantly being improved and enlarged to this day. When the rivers Aire and Calder were originally made navigable up to their present limits, at the very beginning of the eighteenth century, the maximum size of craft carried was thirty tons. A decade ago it was 350 tons, now it is about 500 and likely soon to be raised to 700 tons. These figures just about sum up the story of the *Aire and Calder*'s development and expansion, achieved in the best traditions of private enterprise, and it is encouraging to see how this policy has been continued under nationalized control. The *Aire and Calder* goes from strength to strength.

The *Aire and Calder* is neither canal nor river navigation but more a mixture of the two. To confuse matters further, the waterway is not a single line but a once-extensive system of connected waterways—a system which has contracted to cut off the unprofitable limbs and leave only those which are most useful. However, the question of one-time ownership of other adjoining canals has, since nationalization, become a criterion that is no longer realistic, and so waterways like the *Selby Canal* and the *Dewsbury Cut* of the *Calder and Hebble* are considered separately. The *Aire and Calder Navigation* is therefore taken to be the line from Leeds to Goole, and from Wakefield to Castleford.

In general terms, the navigation is based on the lower reaches of the Aire and Calder rivers, the main line following the Aire down from Leeds, and the Wakefield line following the Calder to join the Aire at Castleford. The enlarged navigation then continues down the Aire to Knottingley, from where a canal cut separates the navigation from the river and leads the former away on a quite separate path to Goole, where there are extensive docks on the edge of the River Ouse. There is little that is attractive in conventional terms about either of these valleys, both of which have been ravaged by coal mining long enough for the landscape to be badly scarred almost everywhere from its after-effects. But the old collieries and power stations that have made such a mark on the Aire valley offer just the conditions for large-scale transport by water, and the *Aire and Calder* has always made the most of this potential.

The navigation starts in the centre of Leeds and follows the course of the river for a short distance past a big basin to Leeds Lock, a basin to which there is a considerable oil traffic using the length of the *Aire and Calder* (and barges which bring the fuel from Immingham Dock on the Humber). Leeds Lock is the start of the first of the canal cuts that run parallel to the river. This cut runs for seven miles through four locks before dropping back into the river at Kippax Locks, a mile short of Castleford. On the way is a waterside power station, bustling with barges. Castleford Junction is a waterway crossroads where the River Aire continues straight on for a weir, the River Calder (the Wakefield Branch) enters from the south-west and the combined navigation enters the Castleford Cut. This is but a short stretch; the navigation rejoins the Aire just to the east and follows the winding, desolate river down to the power stations of Ferrybridge, where it leaves the Aire for the last time to pass through Knottingley (where there is the navigation's only cutting of any substance) and then branches right to leave the coalfields of the Aire valley and enter the flat and—until the recent construction of the M62 motorway—peaceful countryside that stretches to Goole. Long, dead-straight stretches are typical of this waterway, which was

The River Calder in Wakefield.

constructed well after the rest of the navigation. Its opening in 1826 bypassed the lower tidal reaches of the Aire which are very winding, and created the port of Goole. There are only two locks, at Whitley Bridge and Pollington. Both are enormous, and mechanically operated.

The Wakefield Branch is very similar in every way to the Leeds–Castleford route, for the Calder valley along here is a far cry from its course just a few miles upstream. The branch starts in Wakefield on a short spur off the River Calder. It locks down into the river, which turns north at Wakefield Power Station—beside which is the site of the entrance lock into the old *Barnsley Canal*. A large flood lock takes the navigation into a long straight cut, which passes a colliery and then crosses over the River Calder on a low iron aqueduct at Stanley Ferry. Three more locks come after, on the same scale as those on the main line. The new M62 motorway crosses upstream of Altofts Lock, and nearby on a separate cut are the two disused locks that the present one replaces. From here the navigation follows the river course down to Castleford Junction.

Interest in this waterway must lie principally in the activities and traffic that the waterway generates. A look at any one of the locks will reveal to the casual visitor the scale of things on the navigation and the purposeful way in which it is run. Leeds is the upper limit of the main line: the *Aire and Calder* links up with the *Leeds and Liverpool Canal* right in the heart of Leeds and beside the City Station. A steam-driven crane operates at a coal wharf here, and further down at Leeds Lock is the big oil depot; but it is difficult to get at, as is the busy power station at Skelton Grange. Indeed, the whole of the Aire valley from Leeds to Castleford is well off the beaten track and crossed by only one main road. Castleford Junction is, however, very easy to reach by road, lying just off the A656. There is not only the flood lock and toll office, but also Bulholme Lock at the far end of Castleford Cut —which is usually full of barges of all kinds, mainly belonging to Cawoods & Hargreaves who have a boat repair works. Their black and orange craft are a common sight on the *Aire and Calder,* for they and Harker's are the main carriers on the waterway. On the Wakefield Branch, the Stanley Ferry aqueduct is worth a look: it is the only significant aqueduct on the *Aire and Calder,* a strange half-moon of iron struts and girders beside a minor road and easy of access.

But the main focal points of the *Aire and Calder* are probably Knottingley, Ferrybridge and Goole. In Knottingley, the navigation is uncharacteristically narrow, confined by cuttings and complicated by junctions. The old branch down to the River Aire from the east end of the town, with the limited dimensions of Bank Dole Lock, act as a reminder of how the *Aire and Calder* used to be, and how it has grown in the last 150 years. The original bridge by the shipyard at the junction used to be an historic structure, one of the first 'skew' bridges. It has since been replaced by a concrete one. At the west end of the Knottingley Cut is the huge Ferrybridge Lock above which are the three power stations of Ferrybridge. These are difficult to get at except by water—a pity, since one of the *Aire and Calder*'s highlights is tucked away just up the river here: a great hooded hopper into which is discharged a constant flow of coal from barges. The hopper is a modern version of the 'Tom Pudding' system which still operates at Goole. The Ferrybridge installation consists primarily of a one-man operated crane shunting barges along a special water siding. When a barge is in position, a cradle seizes it from underneath, hauls it out of the water and tips it over the wall of the hopper, shooting the coal into the covered bunker behind. Each barge holds up to 160 tons of coal; and a barge can be unloaded every five minutes. These works feed both Ferrybridge B & C stations.

At the east end of the *Aire and Calder* is Goole, a busy inland port at the

Modern coal wharf at Ferrybridge. The large rectangular hoppers are hauled out of the water and tipped into a bunker that feeds the adjacent power stations.

junction with the River Ouse. The elaborately laid out docks, and indeed the whole port, were built from scratch in 1826, and although Goole Docks today owe more to the Ouse than to the canal, there is still plenty of traffic between the two. Among the principal features of the Docks are the old coal hoists, strange iron round-topped gantries which turn the 'Tom Puddings' upside down. Most of the docks are accessible to pedestrians, but it is much easier to explore by boat.

BOATING ON THE AIRE AND CALDER

Taking a pleasure boat along the *Aire and Calder* is a very different experience from boating on the narrow canals. It is in many ways a realistic taste of a typical European waterway of rather modest size. The boater should keep a lookout for commercial traffic—which is no great problem on the broad, winding reaches of the River Aire, or the unrelieved straight miles of the long canal from Knottingley to Goole. But a big tanker cutting a corner or a train of 'Tom Puddings' on a windy day can pose a problem, and the cutting through Knottingley, which has an awkward double bend in the middle, is sufficiently narrow to be the possible source of much embarrass-ment. Traffic lights have been installed at the end of the cutting but they are not often used. Knottingley is in any case a town not to be rushed, for there is another 'blind' double bend at the junction for the River Aire. A busy barge-building and repair yard is sited on both sides of the canal here. Another obvious hazard is the busy Castleford Junction, where there is a flood lock and a toll office right at the confluence of the rivers Aire and Calder; the traffic lights here are used to the full.

By British Standards, the locks on the navigation are massive. They are also mechanized and operated by a lock keeper who sits isolated in his lock control room and deafened by the noise of the electric motors that operate the gates and paddles so quickly. The keepers tend to keep a weather eye open for oncoming craft and usually have the lock ready for whoever arrives, so one's through passage tends to be swift. But the waterway is anyway not particularly heavily locked: there is one every three miles on average. Need-less to say, pleasure boats are expected to give way to commercial traffic at all times.

There are not very many mooring places on the navigation at which a boat can be safely left for any length of time: the wash from a laden barge can wrench a light pleasure boat away from its lines and leave it adrift in the middle of the waterway for the next vessel to run down. But shelter may be sought in the backstream, at the backwater of the weirstream of the river sections, or in the occasional basin or old arm on the canal section. The

Goole Docks.

The old bridge over the River Aire at Ferrybridge which used to carry the M1. It is now closed to traffic and preserved as an Ancient Monument.

Knottingley to Goole Canal is particularly bleak in this respect; but fortunately there is, at the village of Great Heck, a useful basin protected from the canal. This is the base of the South Yorkshire Boat Club and is understandably crowded; but visiting boats can usually be fitted in for a night. There is no pleasure boatyard as such on the *Aire and Calder*.

River Aire and the Selby Canal

This is the route to Selby that was superseded by the opening of the *Knottingley and Goole Canal* in 1826. Originally the navigation did not enter Knottingley at all, for the River Aire itself was the navigation, and the only

Access to water transport was once an important asset. . . .

Typical lock paddle gear on the *Calder and Hebble Navigation.* The big 'handspike' gives excellent leverage on the heavy paddles.

canal cuts were very short lengths at each lock. So the boats kept to the north edge of Knottingley, and the boatmen would have frequented riverside pubs like the 'Sailor's Home'. The lower reaches of the Aire are extremely tortuous, and it is easy to imagine the difficulties and delays it would have caused to boatmen who had always to turn the wind and tide to their advantage. It was partly to avoid the worst of the Aire's route that in 1788 the Aire & Calder Navigation Company built a short cut from West Haddlesey across to the River Ouse. This was the *Selby Canal,* and it had the effect of bringing not only Selby and York but a whole new district much closer to the *Aire and Calder,* with obvious consequent possibilities for the profitable carriage of goods. The *Selby Canal* itself was replaced as the *Aire and Calder's* main line by the construction of the *Knottingley and Goole Canal* fifty years later; but meanwhile, it had given Selby an importance as a river port which it was to sustain until the present day.

The Knottingley–Selby route is still perfectly navigable, although it is little frequented nowadays by commercial craft, for the locks and bridges are far too restricting for modern traffic. But pleasure boats still find it a very useful short cut, which keeps to a mere fourteen miles the distance necessary to navigate on the tidal River Ouse when travelling up to York. The old course of the Aire below the junction with the *Selby Canal* has fallen out of use now, and the old Haddlesey Lock is derelict, its lock cut completely silted up.

The route itself is a quiet and very isolated one. The river at Knottingley is reached by a small lock (Bank Dole Lock) at the end of a short branch of the *Aire and Calder* near the middle of town. The river winds along to Beal, where there is a lock and weir just downstream of the only bridge over the river before the canal at West Haddlesey. The canal itself, which joins the river at right angles in the shadow of Eggborough Power Station, is protected by a flood lock. The only other lock is at the other end of the canal at Selby, where it drops down into the Ouse. The main point of interest today on this undramatic and quiet stretch of waterway are the unusual stone bridges with their long approaches; in the future, it may be the pithead buildings of a new colliery to be built near Selby.

BOATING ON THE RIVER AIRE AND THE SELBY CANAL

This route is used only as a short cut, and few boaters stop for long. The Aire can run fast sometimes, so it is wise when heading downstream towards Beal Lock to keep on the south side of the river towards the lock and not the weir. The lock at Haddlesey is easy to miss in bad light, and it is in-advisable to risk the weir just below the bridge. Selby Lock, and the swing bridge just to the south, are both operated by the lock keeper who lives in the house on the west side of the lock. There are good permanent moorings in the basin by Selby Lock at the junction with the River Ouse.

Calder and Hebble

This waterway extends the *Aire and Calder*'s branch at Wakefield from the low-lying valley of the Aire up an increasingly steep and narrow valley, through long-established manufacturing towns of mellow stone and up to the Pennines. It is a far cry from the bland, despoiled valley of the *Aire and Calder,* with its desolate unrelieved views of coal mines and slag heaps. The

Calder and Hebble is also a much smaller waterway, with locks that are very much a left-over from the last century. It is officially classified as a commercial waterway, a title which it needs but hardly earns. Virtually the only trade on the navigation is coal for a short distance between a colliery and a power station. When and if this last traffic finishes, the *Calder and Hebble* will be a waterway without a role, unless its possibilities for pleasure use are more fully exploited.

If the waterway is hopelessly outdated now for transport, it is by no means short on physical attractions. The Calder valley remains principally an industrial environment that fits so snugly into the scale of the valley it never appears to dominate the landscape; rather, the *Calder and Hebble* scenery is of fine stone warehouses and factories enclosed by rugged hills, and great railway viaducts framing lonely mill chimneys with intimate lanes leading to secluded stone-built farms. It is also a navigation well spiced with pubs.

The *Calder and Hebble* was built essentially as a river navigation based on the Calder but incorporating a few canal cuts (the Hebble has never been more than an insignificant feeder stream). Later, more canal cuts were constructed to keep the navigation out of the river as much as possible—a much-needed benefit, for the Calder was and still is liable to heavy floodwater after rain in the hills. (The floodgates at the head of each lock cut are frequently closed.) Here and there along the canal are short arms of cuts, mostly disused now, that once linked the canal cuts with the mills and factories that would otherwise have been cut off by the new line of the navigation's course. The navigation starts in Wakefield where a short branch connects the *Aire and Calder* to a wide open reach just above a large weir. The navigation heads west, climbing up through Thornes Cut and lock to rejoin the river for a mile before climbing into Broad Cut. A coal staithe on this cut provides what is now the only regular traffic on the canal. Boats load here to carry the coal a few miles upstream to Thornhill Power Station. In between is an aptly industrial area with five locks, and a winding lockless branch to Savile Town Basin, on the edge of Dewsbury. After Thornhill, the navigation slips in and out of the river through Mirfield and Coopers Bridge as far as Brighouse, where the navigation leaves the River Calder for the last time to tackle the remaining six miles alone. This is easily the most attractive part of the navigation, where the hills are near, and the woods dense. The locks come thick and fast, with a flight of three at Salterhebble. By the top lock is a junction; the remains of the Halifax Branch lead off to the right while the main line of the canal turns along the steep hillside to a magnificent terminal basin in Sowerby Bridge.

The lower half of the canal is the more industrial, and part is still used for the coal traffic. This trade is carried in conventional motor barges which can be seen between Horbury and Thornhill. Downstream is Wakefield, where some impressive riverside warehouses still survive.

Shepley Bridge is a pleasant spot at the junction of the Mirfield Cut and the river. There is a lock, a dry dock and a maintenance yard, a good pub across the river and a broad view down the wooded valley.

Brighouse is another interesting place to see the waterway whose passage through the town has been much improved by a little urban landscaping. The main basin is separated from the river by two locks close together and makes a handy mooring site. There is an excellent walk from here along the quiet valley flanked by wooded hills to Elland, passing on the way the mills and the old cut into the river at Brookfoot. Elland has a wharf with an old covered dock and an excellent position looking down the deep valley.

Salterhebble is also an area worthy of exploration. The three locks are squeezed onto a tight corner together, with a little lock house and a junction at the top. The bottom lock has an extraordinary and (to boatmen very tiresome) modern electric guillotine gate arrangement. At the top, the old Halifax Branch is worth exploring, although it is much truncated now. There are some large warehouses and an old dry dock at its present terminus. More excellent examples of this early industrial architecture are to be found at Sowerby Bridge Basin. The scale of these warehouses is a reminder of the former importance of the navigation to the industry and general economy of the area. The *Rochdale Canal* used to continue the line from here, over the hills and down to Manchester via all of ninety-two broad locks. Its remains can be seen at the back of the town.

BOATING ON THE CALDER AND HEBBLE

The *Calder and Hebble* is still staffed by lockkeepers, although these can be scarce on the section upstream of Thornhill Power Station. Many of the detachable ground paddles on this navigation are opened with handspikes (wooden levers about three feet long with a three inch by two inch head) which are chained to the paddle-post but which tend all the same to disappear downstream. Homemade ones are just as good. The main thing to watch for on the navigation is the weirs (some of them are unmarked) which lie in wait for those boats heading downstream which miss the lock cuts. This is not difficult to do when the river is in spate. The navigation in general is not yet very well developed for boating facilities, but people tend to be keen to meet visitors boating up the navigation, and local help is usually forthcoming for those in need. This applies particularly to the two firms established at Wakefield and Brighouse.

Huddersfield Canals

In 1774, a private Act of Parliament authorized Sir John Ramsden—who owned most of Huddersfield at that time—to build a canal connecting the town with the River Calder which had recently been made navigable. This gave Huddersfield a navigable connection with all of the Yorkshire waterway system; and it would have remained a useful but purely local feeder waterway had it not been for the construction twenty years later of a separate canal extending Ramsden's waterway from Huddersfield over the Pennines to Manchester via the *Ashton Canal*. The new waterway was built with narrow locks to correspond to those of the *Ashton*, and became known as the *Huddersfield Narrow Canal;* Ramsden's was called the *Huddersfield Broad Canal.* The narrow one suffered the effects of heavy lockage—seventy-four between Huddersfield and the top of the *Ashton Canal* alone—and it has been legally abandoned since 1944. The *Huddersfield Narrow Canal* has been filled in on its lower lengths, remaining largely intact but unnavigable in the hills. It was chiefly notable for the 5456 yard long Stanedge Tunnel, the longest canal tunnel built in this country.

Sir John Ramsden's canal—the *Huddersfield Broad Canal*—is a much more modest waterway. It is only three and three quarters of a mile long, with nine locks of the same broad dimensions as those of the *Calder and Hebble.* Its course from the latter waterway follows the valley of the River Holme, along which, inevitably, textile mills and factories have established themselves over the years. So the canal passes through a mainly industrial environment (there are some particularly evil smells emanating from one or two of the works) but this is broken up by parkland and the canal is not unattractive. The basin in Huddersfield is alive with pleasure boats generated by a young boating business. Nearby is a very unusual lift bridge with a complicated arrangement of pulleys and counterweights; despite all this gear, the bridge is still very stiff to operate.

BOATING ON THE HUDDERSFIELD BROAD CANAL

The canal starts at Coopers Bridge, on the river behind the *Calder and Hebble's* lock cut and just above a weir. So it is worth taking care on the

Lift bridge on the *Huddersfield Broad Canal.*

approach; and since the canal is not heavily used, it is also worth warning British Waterways staff of your intentions so that they can ensure that the canal is well filled with water. The often-leaky lock gates tend to leave the water level in the pounds rather low.

Sheffield and South Yorkshire Navigation

The *Sheffield and South Yorkshire* is based upon the River Don—a waterway that has been navigable for centuries. So while the navigation alternates between following the river course and branching out by itself along artificial cuts, it remains at all times within the confines of the Don Valley, one of the most heavily industrialized in all Yorkshire. Sheffield, Rotherham, Doncaster and many smaller towns have grown up along the banks of the Don, and their expansion sideways has created an almost continuous ribbon of industry for twenty-five miles from Sheffield to east of Doncaster. Not surprisingly, the river water is heavily polluted, although even here anglers insist on trying their luck. At the top end especially, industry is on a grand scale: lighter and more modern industry scarcely gets a look in. There are steelworks, heavy engineering factories, coal mines, slag heaps and power stations. Everywhere there are railways, crossing the valley or running along it; and everywhere is the sound and smell of things being produced, used, shifted or changed. It is not a glamorous place, but it is here that real wealth is made, and a journey along the navigation gives a unique insight into this process.

North-east of Doncaster, the waterway is very different: both industry and the River Don are left behind as the navigation abandons the river for a separate, artificial route through open country to the River Trent at Keadby. Meantime, the navigation is full of surprises—like the little old brick bridges in the heart of Sheffield, as old as the canal itself, and the startling sight of Conisbrough Castle a little further down, or the lovely, grand wooded section of waterway either side of Sprotborough. But perhaps the biggest surprise of all is that, in view of its unique geographical position as a direct link between the Don valley and the Humber ports, the waterway's freight potential is so completely wasted. The story of the lack of modernization of this navigation is a sore point, a long saga of missed opportunities, lack of initiative and utter stinginess and lack of interest on the part of one Government after another. The upshot of it all is that the waterway, although splendidly up to date in certain places, is still completely unsuitable for contemporary traffic. Improvements have been made piece-meal for over a hundred years, and while it is good to rebuild locks to double their original capacity, it is pointless unless all the locks are left with increased capacity. The latest scheme for thoroughgoing improvement and enlargement of the waterway has been repeatedly turned down by British governments ever since it was first submitted by the British Waterways Board in 1966, even though it was a project designed to cost less than £3 million. So the *Sheffield and South Yorkshire Navigation* remains largely as it was a century ago, and while the nearby *Aire and Calder* goes from strength to strength in amounts of goods carried, the *Sheffield and South Yorkshire Navigation* handles less and less. However, the Board has never ceased to harrass successive Ministers about the improvement scheme, and if it is ever carried through, it will be largely due to the Board's persistence.

Outdated—the *Sheffield and South Yorkshire Navigation* at Tinsley. Modernization of this waterway has been repeatedly shelved by successive governments.

The *Sheffield and South Yorkshire Navigation* begins its journey in the heart of Sheffield itself, its terminal basin hidden away behind a timber yard. For the first four miles, the waterway (which constitutes the *Sheffield Canal*) traverses the built-up hillside on the south bank of the Don valley overlooking the agglomeration of endless steel mills below. At Tinsley, a flight of eleven locks takes it down into the River Don. This top stretch between Sheffield and the river was added fairly late in terms of the River Don's age as a transport route: it was not opened until 1819, and only then at the insistence—and expense—of the steel cutlery interests in the city. The canal has never been significantly modified since then, and the eleven locks (there were originally twelve), in addition to the necessity of pumping water up to feed the canal, have always made the *Sheffield Canal* something of a 'loss-leader' for the navigation as a whole; and now that there is no more commercial traffic at the basin, its future continues to look uncertain.

From Tinsley onwards the navigation begins to flirt with the river itself, first joining it, then suddenly leaving for a distance before dropping back to rejoin it. At each of these cuts there is, as on most river navigations, a flood lock or at least a pair of 'flood gates' to protect the cut from any river flooding. In this fashion, the navigation continues through the heavily industrial reaches of the Don valley. The waterway is not particularly heavily locked, but the limited size of the locks is noticeable, and so are the often narrow and difficult bends of the channel; these are both hallmarks of a waterway quite out of touch with any modern scale of transport. At Rotherham, however, is a large and active canalside depot, and from here downwards, indeed, the waterway is used to a certain extent by commercial traffic.

A further stretch of winding river course is followed by a long canal section through Swinton and Mexborough. On the way, the waterway exchanges the steel mills for more and more coal mines and their attendant slag heaps. A flight of four locks at Swinton indicates the south end of the former *Dearne and Dove Canal,* now disused. The proud Norman keep of Conisbrough Castle overlooks the river valley that runs deep and wooded down to Sprotborough, where the unexpectedly handsome view of the lock is complemented by the two viaducts that sweep across the valley further down. Doncaster brings a temporary return of industrial flavour to the canal, and the barges that crowd the wharf at the power station are pointers to the navigation's increased use from here downwards. The trains of 'Tom Pudding' compartment boats take the coal down from here to Goole, at the eastern end of the *Aire and Calder Navigation.* The bigger (215 feet long) of the two locks at Long Sandall (built in 1959) is another hint of how things might have been all the way up this waterway if official short-sightedness had not prevailed over informed advice on the commercial development of waterways. Meanwhile, the canal reaches open country at last and arrives at a finely angled junction. The main line of the *Sheffield and South Yorkshire* bears east here: at Stainforth it used to lock down into the River Don to run down on the tide to Goole, but this exit has now been closed and all craft use the *Stainforth and Keadby Canal,* which keeps well to the south and passes through a strangely empty region of dead flat fenland. With a busy railway on the north bank, and with bridge keepers in attendance all the way to supervise the many small opening bridges, the canal negotiates a sliding railway bridge and arrives at Keadby, the junction with the River Trent.

The *Sheffield and South Yorkshire Navigation* upstream of Doncaster, at Sprotborough Lock.

THINGS TO SEE ON THE SHEFFIELD AND SOUTH YORKSHIRE NAVIGATION

Being a largely industrial and often inaccessible navigation, the *Sheffield and South Yorkshire* is in many ways difficult to appreciate at all except by boat. This applies particularly to the section from Sheffield down to Barnby Dun, which is principally an industrial waterway based on the Don, and interest in this upper section tends to lie mainly in the traffic which uses it. Specific points of activity along the navigation include Doncaster, where 'Tom Puddings' are marshalled and loaded at the wharf by the town centre, while conventional motor barges discharge coal at the nearby power station, and Rotherham Depot (BACAT barges use this as a terminal).

Swinton Junction is worth a visit: the bottom four locks of the *Dearne*

A boatyard at Swinton Junction on the *Sheffield and South Yorkshire Navigation*.

and Dove Canal are here, still in use for a small traffic to a nearby glassworks. There is also Waddington's boatyard at this point—one of the old carrying firms that has soldiered on with trading along the navigation. Boats have been built here for many, many years, but now most of them are here to be scrapped.

Sheffield Basin was once another focal point on the navigation, but its fine range of canal warehouses is now largely unused, for the restricted dimensions of Tinsley Locks have finally stifled any traffic up to the terminus. The basin is now occupied by pleasure boats, but at least it is used.

The other end of the navigation is a great contrast to the valley of the Don. Most of the traffic turns off the main line of the *Sheffield and South Yorkshire* to use the *New Junction Canal* as a route to Goole, leaving the line from Bramwith Junction to Keadby relatively deserted. It is a very different waterway indeed to the navigation between Sheffield and Doncaster. The route passes through the flat, featureless countryside that lies to the west of the River Trent.

Sheffield Basin.

The remains of the old lock and basin at Stainforth are also worth a look, as is Thorne, an attractive town where pleasure boats abound, and Keadby, the small riverside wharf on the Trent beside the big iron bridge across the river.

BOATING ON THE SHEFFIELD AND SOUTH YORKSHIRE

The very idea of taking a pleasure boat up the *Sheffield and South Yorkshire Navigation* would strike many people as odd, but those who wish to broaden their canalling experience will find it interesting—and it is not so much of an unexplored backwater as one might imagine. The *Stainforth and Keadby Canal* has been a good mooring for river-going boats for a long time, with a heavy concentration of boats and facilities at Thorne. The back of Long Sandall Lock is also a popular mooring place but the industrial valley of the Don is not particularly conducive to cruising, and the next outpost of pleasure boating is right up at Sheffield Basin; in between is the bulk of the navigation proper. On the river there is, of course, for every lock a weir nearby; these are not always well marked or fenced, so it is sensible to take

133

care if the river is in flood. Most of the locks are attended by lock keepers, which makes the going easy. The flood locks or gates at the head of the lock cuts are not attended, and they are normally open except in flood conditions.

New Junction Canal

No more than a link route connecting the *Sheffield and South Yorkshire Navigation* with the *Aire and Calder Navigation,* this is five and a half miles long, and it is dead straight. The canal's opening in 1905 signalled the virtual end of the River Don as a part of the area's transport network, for the new cut gave direct access for coal from the Doncaster area to the docks at Goole.

The *New Junction* is as far removed from the average English narrow canal as one can imagine. In its appearance and in its operation the canal owes more to Holland than to this country. Not that it is particularly big; as with most ordinary Dutch waterways, it is the depth that counts more than any great width. But the canal is, again like the Dutch canals, consciously operated as a service to the floating customers that continue to use it—to everyone's profit. Thus at each of the six swing bridges on the canal (the only other bridge is a fixed foot bridge) there is a house; at each house lives a bridge keeper, and each time a boat appears the bridge keeper turns out to open the bridge with a capstan that stands in his garden. A simple and painless procedure perhaps, but the service is a reflection of the high level of traffic that this very business-like canal still enjoys.

There is just one lock—Sykehouse Lock. This is long (195 feet) by most standards, but not long enough for the trains of 'Tom Puddings' that have to unhitch and split up in order to get through; fortunately this is the last lock before Goole. There are also two small aqueducts, one at each end of the canal. At the north end, the junction with the *Aire and Calder* main line, a minor aqueduct carries the *New Junction* over the River Went. The Went is a tributary of the River Don which itself passes under the south end of the canal. The latter is carried over on an iron structure, and great guillotine gates rear up over the navigation. Most of the way the canal is carried on a slight embankment through the landscape, and since the latter is flat, quiet grazing country, the canal enjoys a pleasant situation. An odd effect is produced by the cylindrical towers of the power station that breathe over the few southern miles of the canal. Their presence combines with the flatness of the land and the absolutely regularity of the canal's course to give an oddly geometrical feeling to the area.

6

Merseyside Waterways

WEAVER NAVIGATION
BRIDGEWATER CANAL
MANCHESTER SHIP CANAL

Whereas in the Midlands the early canals were built to fit narrow-beamed craft, those on Merseyside were built broad from the outset to accommodate the river barges already trading in the region. Thus the Duke of Bridgewater's canal was no water by way, but was conceived as a principal new trunk route, and still stands out today as a great contrast to the relatively small-scale canals that followed. The *Manchester Ship Canal* is worlds removed in scale from the Bridgewater, but it is very much its direct descendant in many ways. Between these two extremes of scale is the *Weaver Navigation,* which can take 650-ton craft. The *Weaver* is well-used commercially but in spite of its secluded reaches is little known to pleasure boaters. The *Weaver* joins the *Manchester Ship Canal* at the busy Weston Point Docks.

Weaver Navigation

The *Weaver* is one of England's least-known commercial navigations. Only twenty miles or so of navigable waters, and tucked away for much of this distance in a thickly wooded, road-free valley, the *Weaver* is yet an important trading waterway, despite its being a somewhat remote limb of the connected system. Those who glimpse the river from a train, or a car crossing one or other of the big bridges that straddle the valley, are often surprised to see large sea-going ships steaming along beneath them or negotiating one of the enormous locks.

The *Weaver* is in fact one of the early river navigations in the Merseyside area; it was made navigable as a controlled, non-tidal river between 1730 and 1732, shortly after the Mersey itself was made navigable up to Manchester. The scale of the *Weaver Navigation* was very different from what it

The *Trent and Mersey*'s access to the *Weaver Navigation*—via Anderton Lift.

is today: it had eleven locks, all barge size, in comparison with the five great ship locks of today. But the river was a busy one, for its prime purpose was to carry salt from Northwich and Winsford to the Mersey and on, while clay and coal for the Potteries provided return cargoes. Later, the emphasis switched from salt to chemicals, and the continual enlargement of the navigation channel and the locks has helped to maintain the *Weaver's* prosperity. However, the river does not carry as much as it might do, and there is little trade above Northwich today. A recent ambitious scheme to develop a modern barge terminal at Winsford has been thwarted several times by successive governments—to whom the idea of water transport as a viable proposition has always seemed the suggestion of mere dreamers.

The river itself is about fifty miles long. It rises only about twelve miles south-west of Northwich, circling round through Audlem and Nantwich before passing under the Middlewich branch of the *Shropshire Union Canal.* Thus far, the *Weaver* is only a large stream like hundreds of others in the country. But just south of Winsford the scene changes completely as the stream broadens out into the chain of three 'flashes'—lakes caused by local brine-pumping subsidence. At Winsford Bridge is the official head of the navigation proper, and although it's an unpromising start (with high banks apparently little more than the edges of huge waste heaps, the scars of the salt extraction industry), this is soon left behind and the valley becomes narrow and wooded. The glorious *Vale Royal Cut,* green, secluded and peaceful, is more typical of the *Weaver,* as are the elegant stone railway viaducts that soar high over the river to allow plenty of headroom for ships. There are two on the south side of Northwich, and the best of all are at Pickerings Cut. Roads cross the river mainly by great swing bridges; but no roads follow the river itself, and the valley is for most of its length completely free of road traffic—a remarkable feature in a region as crowded as this.

Northwich has always been the main centre of activity on the *Weaver*, and although the barge-building trade has contracted considerably, there is still plenty going on along the river banks. North of the town is another industrial area, based on a big chemical works at Winnington. The river's principal trade is conducted here, as various bulk chemicals are loaded into large coasters for export. Opposite the main wharf is the *Trent and Mersey Canal*, joined to the *Weaver* by the famous Anderton Lift. This is the way most pleasure boats reach the river, especially since the closure of the little *Runcorn and Weston Canal* removed the *Weaver's* link with the *Bridgewater Canal*. Nowadays, the only other link is with the *Manchester Ship Canal*.

Below Winnington, the *Trent and Mersey* follows the hillside overlooking the *Weaver*. The latter dodges from river channel into artificial cut and back again. There are only two sets of locks, at Saltersford and Dutton; others, at Acton Bridge, Pickerings and Frodsham, have long since disappeared. Their absence consolidates the river's sense of remoteness, wandering as it does through a hidden corridor of green, free of human influences except for the occasional farm or old lock house along the banks. To find a small ship picking its way along this most inland-looking valley is indeed a bizarre experience.

At Frodsham the Weaver valley opens out as it reaches Merseyside, and soon resigns itself to the world of vast chemical installations and abrasive smells that announces the *Manchester Ship Canal*. The old lock down towards the Mersey can still be seen, its limited dimensions bearing witness to the improvements to the river since it fell out of use. The navigation's present exit to the Mersey is along the broad *Weston Canal*, built in 1806 along with the locks at Weston Point. There is a side lock into the *Ship Canal* which runs right beside the Mersey here, and another lock down from the docks.

BOATING ON THE RIVER WEAVER

Because of the *Weaver's* relative inaccessibility by road and the absence of any formal towpath for much of the way, it is difficult to get any realistic impression of the navigation except by boat. It is of course excellent boating water, with plenty of space and generally very pleasing surroundings. The five locks are all extremely large and all operated by lock keepers—which means that opening times at weekends are limited. (This applies also to the swing bridges which are all manned, although many of the big ones allow

Worsley Delph on the *Bridgewater Canal,* where the canal age began. Under the bridge is the entrance to the Duke of Bridgewater's mines (no longer operative). The water here is bright orange from mine seepage.

ample headroom for pleasure boats without needing to be opened.) Below Northwich, the river's use by coasters sometimes extends to night-time as well, so pleasure boatmen should be careful about tying up overnight. It is well worth exploring the top end of the river and penetrating at least to Winsford Bottom Flash. The edges are very shallow, so it is best to be able to anchor out in the middle of the lake. Also worth exploring are Weston Point Docks, the nearby remains of the *Runcorn and Weston Canal,* the old Frodsham Cut, and the other traces of now superseded locks.

Halfway along the river is the Anderton Lift. Now a century old, the lift is the vital link with the *Trent and Mersey Canal.* The lift keeper, who works the same hours as the lock and bridge men, makes a charge for the use of this extraordinary and unique piece of machinery.

The Barton Swing Aqueduct—in action here—carries the *Bridgewater Canal* over the *Manchester Ship Canal.*

Bridgewater Canal

Although it was by no means the first canal in Britain, the *Bridgewater* was of cardinal importance to the great industrial revolution that got under way in the second half of the eighteenth century. The hitherto untried technique of building a navigation that was in no way constrained by or dependent upon any given river valley, and the revolutionary idea of bridging not only roads but a substantial river by a large canal-containing structure, were both shown to work in practical and commercial terms. The proven utility and profitability of such an approach was an eye-opener for engineers, entre-preneurs and manufacturers throughout the land who had been accustomed to seeing the advantages of water transport over pack-horses confined always to river valleys. It was the possibilities suddenly revealed by the *Bridgewater Canal's* success that triggered off the great age of canal building in Britain.

The story of the project is well known—the young Duke of Bridgewater's observation of water transport developments in Europe; his plan for a direct water highway between his coal mines at Worsley to the consumers in nearby Manchester; the building of a level canal by the Duke's men James Brindley and John Gilbert; the scepticism and disbelief from all quarters that greeted the proposal for a masonry aqueduct to carry the canal and its boats nearly forty feet above the River Irwell at Barton; Brindley's 'puddled' clay to seal the canal bed; and eventually the recognition and acclaim for this bold and exciting achievement.

The Duke of Bridgewater's canal was indeed something totally new in this country, and although it is therefore now older than almost all other canals in Britain, it is by no means the primitive kind of waterway that one might expect to find. It is a tribute to the canal's builders that regular commercial traffic on the *Bridgewater* has outlasted that on virtually all narrow canals and even on the *Grand Union*. The last big, regular traffic was imported grain for Kellogg's Trafford Park works. This used to come by ship to Manchester, to be discharged into *Bridgewater Canal* boats which then came up Hulme Lock and round by the *Bridgewater Canal* to Trafford Park. Now it comes from Liverpool, entirely by road.

The original *Bridgewater Canal,* opened in 1761, ran from the entrance to the Duke's coal mines at Worsley over the River Irwell at Barton and on into Manchester. Soon afterwards, he had a second and longer line built, from the canal at Stretford right through to Runcorn, descending by ten locks to the River Mersey; a short branch at Preston Brook made a junction with the new *Trent and Mersey Canal.* Later, in 1795, the Duke acquired Parliamentary authority to build a canal north-west from Worsley to meet the new *Leeds and Liverpool Canal*'s branch to Leigh.

Meanwhile, back at the Worsley mines, a logical development of the system had been to take canals inside the shallow mines, and a remarkable and extensive network of narrow canal tunnels was developed on various levels, connected by inclined planes enabling special tub boats to fetch the coal straight from the faces. In fact, the *Bridgewater Canal* was not only a vital prototype canal—it was also a profitable, self-contained canal system as well as becoming part of a through route between the north-west Midlands and the *Leeds and Liverpool Canal.*

The opening of the *Manchester Ship Canal* in 1894 did not affect the *Bridgewater*'s status as a through route, for the only place where the *Ship Canal* intersects the older waterway is at Barton, where the original three-arched aqueduct was replaced by an equally revolutionary swing aqueduct to allow ships unlimited headroom. But a more recent and less welcome modification to the canal has been at Runcorn, where the canal originally locked down into the Mersey, and later into the *Runcorn and Weston Canal* and the *Manchester Ship Canal*. Today, the *Bridgewater Canal* in Runcorn is a cul-de-sac. The Runcorn locks were closed in 1966 and unceremoniously filled with concrete.

THE COURSE OF THE CANAL

The canal from Manchester to Runcorn, nowadays regarded as the main line, runs roughly east–west along the south side of the Mersey valley. The scenery is typical of lowland Lancashire–Cheshire; unfussy, uncompromising and in stretches overwhelmingly industrial, although many miles are as unspoilt as can be expected on Merseyside. There are no locks along the way, for the canal tends to follow the contours, but the numerous minor aqueducts and the size of the bridges and embankments are a reminder of the very ambitious scale of the original undertaking, which was clearly not just designed for the narrowboats that became the standard gauge on later canals.

The eastern terminus of the *Bridgewater Canal* is the Castlefield Basin in Manchester, now used only as a base for maintenance craft. Nearby is the junction with the *Rochdale Canal,* and the very brief Hulme Locks Branch which nowadays drops through only a single deep lock into the upper reaches of the Irwell and thus into the *Ship Canal*. A little to the west, the *Bridgewater Canal* overlooks the terminal docks of the *Ship Canal*. From Stretford to Altrincham the navigation runs south-west, dogged by a busy railway line before shrugging off the outskirts of Greater Manchester. The contours of a shallow hillside lead it west to the villages of Dunham and Bollington. The great embankment here was the scene of a disastrous breach in 1971 and the aqueduct over the River Bollin has been completely rebuilt. Further west, past Warrington, is Preston Brook. Now overshadowed by the M56 motorway, the canal meets the *Trent and Mersey* via a short branch, then doubles back over the railway tracks to an interesting contrast between an old interchange basin and a huge new twelve-acre, 450-berth marina for pleasure boats. The waterway from here to Runcorn, once entirely rural and giving excellent views over the River Mersey, is now flanked by the box-like dwellings of Runcorn New Town—a sorry fate.

The other main limb of the *Bridgewater Canal* is from Stretford to Leigh, taking in the original line from Worsley. The first section of this route to Worsley is a rather grim industrial stretch; but beyond Worsley the canal runs along the bleak northern edge of Chat Moss to the cotton mills of Leigh. The two principal features on this long pound are the old terminus of the canal at Worsley and, of course, the Barton Swing Aqueduct. The former, an attractive and sheltered spot, still features the tunnel entrance to the mines.

141

'The Packet House', a canal inn nearby, recalls the passenger service to Manchester by boat initiated by the old Duke.

The Barton Swing Aqueduct attracts the curious from all over the world. It is a unique piece of machinery and a worthy successor to the original aqueduct. The present bridge is essentially a steel tank 234 feet long, pivoting on an island in the centre of the *Ship Canal*. The water in the tank is six feet deep and eighteen feet wide, with an elevated towpath over the water. To make way for traffic on the *Ship Canal,* the whole tank is simply sealed and swivelled by ninety degrees, still full of water. Its rigid strength must be immense to carry the 1500 tons that the whole giant tankful weighs.

BOATING ON THE BRIDGEWATER CANAL

Thanks to the Duke of Bridgewater and his men, the remarkably generous limiting dimensions of this canal, and the very long single level on which it was built, make the *Bridgewater* a very straightforward canal to navigate. And following the run-down of commercial traffic, the canal's value as a pleasure-boating waterway has been more and more widely appreciated by the boating fraternity. There are now half a dozen clubs on the canal and an increasing number of boatyards—notably the big new development at Preston Brook. The canal is owned and administered by the Manchester Ship Canal Company (who bought out the canal before building their own waterway). The Company naturally levies licence dues on pleasure boats based on the canal, but under a reciprocal arrangement, visiting boats licensed on British Waterways Board canals are allowed to pass along the canal free of charge.

Manchester Ship Canal

The sight of a big ocean-going ship creeping along through crowded Merseyside, towering above factories and housing estates and squeezing under the tall, fixed bridges, is indeed an impressive one. Ship canals in Britain are few and far between, and the *Manchester Ship Canal* is far bigger than any of the others. Indeed, this canal is so far removed in scale from all other canals in Britain that it seems unrealistic in a way to classify it with the others; it differs only in scale, however, and not in principle. It is not difficult to draw a parallel between the *Ship Canal* and the old *Bridgewater*

Barges beside the *Manchester Ship Canal* at Ellesmere Port.

Canal, not only for their similar route but for the engineering progress that each represented in its day and the economic benefit that each conferred on the city of Manchester.

The *Manchester Ship Canal* was first successfully promoted in the early 1880s as a means of turning the great manufacturing conurbation of Manchester into a seaport on a par with Liverpool. The scheme came in for more than its fair share of scepticism—as indeed had the Duke of Bridgewater's scheme 130 years previously—but the *Ship Canal* was no private venture, it was a great public enterprise financed by the City of Manchester itself.

Construction of the *Ship Canal* involved buying out the *Bridgewater Canal* entirely—for £1·7 million—and obliterating much of the *Irwell Navigation,* most of whose upper reaches were absorbed into the new seaway. Around Irlam the Mersey too became part of the new canal; but further downstream the Mersey was left to go its own way to the sea, becoming a very wide and shallow estuary. The *Ship Canal* was made as a completely independent canal beside the estuary, through Runcorn and Ellesmere Port to Eastham, almost opposite Liverpool.

It was opened by Queen Victoria in 1894 and has been a tremendous success ever since, having quickly evolved into a single linear port right along Merseyside. Oil refineries, chemical plants and much other heavy industry have all grown up along its banks. The annual tonnage handled here is enormous and includes ten million tons of oil and oil products. Not surprisingly, its waters are not the purest. They have been known to become so polluted with oil and chemical spillage that on at least one occasion a considerable length of the waterway's entire surface has caught fire. The canal handles ships up to 12,500 tons to Manchester. There are five sets of locks on the canal and four of these are paired, the bigger lock of each pair measuring a vast 600 feet by sixty-five feet, while Eastham Locks are a set of three of which the biggest is eighty feet wide.

Eastham is the western entrance to the canal up from the River Mersey. From here it follows its stately way right along the river bank past Ellesmere Port and Stanlow, where the principal oil wharves are located. Ships displacing up to 15,000 tons can reach this far. Further east, the canal broadens out where it receives the waters of the River Weaver. Around the corner are Weston Point Docks—the flourishing terminus of the *Weaver Navigation*— and Runcorn Gap. The old ferry and the transporter bridge have disappeared from Runcorn now, and both the Mersey and the canal are spanned by conventional road and rail bridges. These are the first bridges across the canal since Eastham, and it is an interesting fact that on this of all waterways fixed bridges should ever have been allowed. The headroom they offer is considerable, the lowest being about seventy-five feet above the water, but this is not enough to clear the masts of many ships using the waterway. As a result, many vessels have to be 'undressed' by a crane at Eastham of any upperworks that would foul the bridges. (Ships that use the navigation regularly have telescopic masts.) The reason for the inconvenient existence of fixed bridges over the *Ship Canal* may simply be that the waterway arrived so late, well after the completion of many of the railways in the area, that it was impossible for the newcomer to take priority over these and, by swing bridges, disrupt the railway companies' timetables.

East of Runcorn the *Ship Canal* continues to thrust along the tapering

Mersey estuary, the canal itself narrowing to a cutting at Warrington not wide enough for two ships to pass. East of Latchford Locks is the M6 motorway viaduct, followed by a short stretch of open landscape before the onset of factories and wharves and another set of locks at Irlam. As the canal enters the outskirts of Manchester it meets the two swing bridges at Barton (one of them being the aqueduct) and then a final set of locks before it terminates at extensive docks in Salford. The River Irwell, whose waters are absorbed by the *Ship Canal* east of Rixton, near Warrington, continues as a navigable channel for a short distance upstream, providing a link to Hulme Lock and thus to the nearby *Bridgewater Canal*. The river used also to give access to the *Manchester, Bolton and Bury Canal* and to the *Manchester and Salford Junction Canal,* but these are now both abandoned.

BOATING ON THE SHIP CANAL

In terms of the inland cruising network, the *Ship Canal* is not a particularly important route—nor was it ever designed to be. The *Bridgewater Canal* continues to act as the all-important link from north to south as well as forming part of the *Cheshire Ring.* The *Ship Canal,* on the other hand, although it joins up the northern ends of the *Shropshire Union Canal* and the *Weaver Navigation*—with a connection to the *Bridgewater Canal* in Manchester —can be regarded as irrelevant to the inland network, although it provided a useful diversion during the two-year stoppage on the *Bridgewater* following the breach in 1971.

The canal anyway is in no sense geared to pleasure craft, and their passage along the waterway is very strictly controlled. The *Ship Canal* authorities insist on guarantees of boats being entirely seaworthy and adequately equipped with charts, tidetables, anchor cable and so on. Boatowners should contact the authorities well in advance to find out the requirements and the procedure. An easier and perhaps more interesting way to navigate the canal would of course be by ship, getting the full benefit of the big locks and the swing bridges. It is possible in summer to join a day excursion on a charter vessel running all the way from Manchester to Liverpool—or vice versa—and then to return by train.

7

North Lancashire Canals

LEEDS AND LIVERPOOL CANAL

LANCASTER CANAL

These are two northern waterways which had much in common, and actually shared the same route to cut their costs. The closure of the link between them has left the *Lancaster* accessible only by sea.

Leeds and Liverpool Canal

One hundred and twenty-seven miles long from end to end, the *Leeds and Liverpool* was the longest single canal ever opened in Britain. It was fifty years building and cost over one million pounds, an unheard-of price for a canal at that time; it must have seemed a daunting figure for a daunting task. Its course, joining the Mersey to the rivers of south Yorkshire, involved crossing the Pennines, something which had never before been attempted, and yet one which presented an inescapable challenge to canal builders if east–west traffic was not always to be confined to pack horses, or the long water route via the *Trent and Mersey Canal,* skirting round the southern edge of the Pennine ridge a hundred miles to the south. And so it was built, and remains one of England's grandest canals, the epitome of a former working canal; yet the usefulness of this, the first trans-Pennine waterway (and a wide one at that) connecting the prime manufacturing towns and the principal river basins of Lancashire and Yorkshire, was undermined by the problem of the heavy lockage required. The canal was never intended to be any kind of large-scale sea-to-sea waterway, like the *Caledonian Canal* in Scotland, and since its earliest days it has in fact proved more useful as a carrier of local traffic than as a transporter of long-distance merchandise.

The canal was first authorized in 1770, and construction started briskly from end to end, thus extending what was already navigable. In this way, progress along Airedale from Leeds soon placed a whole string of Yorkshire manufacturing towns (Shipley, Bingley, Keighley and Skipton) firmly on the waterway map, linking them directly to the *Aire and Calder Navigation*

The *Leeds and Liverpool Canal* near Shipley.

at Leeds. At the other end, meanwhile, the old *River Douglas Navigation* was bought out and a considerable length of canal constructed, joining Liverpool to Wigan. All of this was achieved within a few years of the canal's authorization. But after this, progress ground to a halt as the promoters struggled to find extra money to finance construction of the more difficult section in the middle. In 1790, a drastic revision of the route added many miles and two tunnels to the total length.

By incorporating the fast-expanding towns of the Calder Valley into the route, it seemed that the necessary finance would be found. Even so, the last link in the Liverpool–Leeds chain was not forged until 1816, and this only by a useful deal with the *Lancaster Canal* by which the former used the latter's course for ten miles, from Wigan Top Lock to Johnsons Hillock. Completion of the Leigh Branch in 1821 gave the *Leeds and Liverpool* access to the *Bridgewater Canal* and thus a line to the south.

This canal is the only remaining Lancashire–Yorkshire waterway. The other two—the *Huddersfield Narrow* and the *Rochdale*—are defunct; somehow the *Leeds and Liverpool* has managed to survive intact, with all its branches save the Walton Summit. There is no longer any commercial traffic on the canal—the last main traffics were mainly coal to waterside power stations along the route—but it is gradually gaining in popularity as a pleasure boating waterway. Much of this is strictly local, but the canal is very valuable as a through route, firstly between two seas and secondly as part of an enormous circuit of waterways extending right down to southern Derbyshire. More and more boat owners are tackling this round trip, in several stages. It is an illuminating experience, especially for the Southerner, for a trip along the *Leeds and Liverpool* provides not only an interesting contrast to canals elsewhere in the country, but also a real view of the face of the working North.

147

The scale of all these industrial towns tucked away in these northern valleys for over two centuries, the important role of an adequate water supply for the textile mills, and the later importance of first the canal and later the railways for transport of materials, are all brought home forcibly to anyone travelling slowly through the region. Yet although, inevitably, it is the towns along the canal that leave the strongest impression on the traveller, it is by no means a gloomy industrial waterway and takes in a whole gamut of landscapes.

THE COURSE OF THE CANAL

LIVERPOOL TO WIGAN: Much of this length is a single long pound through the northern plains of Lancashire. In Liverpool itself is one of the grimmest lengths of urban canal in the land, with miles of dreary housing estates built on its banks. From most of these houses the canal is fenced off in an entirely unsuccessful official attempt to keep small boys from playing on the towpath, falling in the cut and drowning. With the same purpose in mind, the swing bridges are all fortified with wire netting and padlocks, and there is even wire overhead, in the form of a major electricity transmission line whose pylons straddle the canal bed. Nothing remains to be seen of the original terminal basins in Liverpool nor indeed any worthwhile canal architecture; but unfortunately the Liverpool end is important to inland navigators as the four locks of the Stanley Dock branch, near the canal's present terminus, give access for full-length (seventy feet) boats to Liverpool Docks and thus to the Mersey estuary.

From the outskirts of Liverpool to the beginnings of Wigan is a journey through the flat and rather featureless farmlands of Lancashire, then up the little valley of the River Douglas into Wigan itself. A junction near the canal town of Burscough drops a seven lock branch north to the tidal waters of the River Douglas, the Ribble and the sea—surely one of the least-known

The *Leeds and Liverpool Canal* at Wigan.

Bingley 'Five-Rise'—a twelve feet drop per staircase lock.

and least spoilt backwaters of the canal system. From Wigan, a seven and a quarter mile branch leads to the *Bridgewater Canal* at Leigh. The Leigh Branch runs through a breathtakingly bleak area laid waste by mining operations. (Fortunately there are only two locks between Wigan and Manchester, so boats can hurry on through.) At one of the collieries on the Leigh Branch originated the very last regular traffic carried on the *Leeds and Liverpool Canal*— coal for the power station at the canal junction in Wigan.

WIGAN TO BARROWFORD: The canal at Wigan is notable more for the great flight of locks than anything else: there are twenty-one in a single flight. The local lads are always pleased to see boats travelling up the locks on a hot summer's day, for it helps to top up these wide deep locks: with their built-in lock-ladder up the side, these make ideal and quite illegal swimming pools. From the top of the flight a single level extends along the western edge of the Pennines, past great mills at Chorley to the foot of six locks at Johnsons Hillock, where the old Walton Summit Branch used to continue the *Lancaster Canal*'s canal-and-tramway route to Preston. Here the main line of the *Leeds and Liverpool* turns away inland to broach the Pennines. But first it has to negotiate the unglamourous belt of sturdy and uncompromising Lancashire cotton towns—Blackburn, Burnley, Nelson and other smaller places. The canal once meant much to these towns and traces of its former relationship can be seen everywhere in the old wharves, with their warehouses, stables, cranes and other paraphernalia. General economic circumstances have left these long-established industrial areas relatively unchanged, particularly in the once-thriving environs of the canal as it winds along through the middle of all this; the flavour of the nineteenth century is still strong.

The waterway dodges this way and that along the hillside, in and out of these uniformly built-up areas, but it is by no means stifled by its surroundings, for beyond the huddled terraces of cottages and the mill chimneys is rugged upland scenery. Notable features on this stretch include a tunnel, and a long embankment across Burnley. The latter is often described as one of the 'wonders of the waterways', but in truth it is hardly spectacular—even if the raising of it represented a great feature of engineering at that time.

BARROWFORD TO BINGLEY: This is a dramatic contrast to the section just to the west. At Barrowford, seven locks hoist the navigation out of industrial Lancashire and—via a long tunnel and a gap in the hills—over into upland Yorkshire. Soon the canal is dodging along between rounded hilltops before dropping to the very head of Airedale. At the back of the stone-built villages rise the Yorkshire Dales. There are locks tucked away

The paddle opening on a modern *Leeds and Liverpool* gate.

in the trees at Bank Newton, and then a rash of swing bridges—well over twenty between Gargrave and Bingley. Skipton is an excellent place to see the canal, and there is much 'messing about in boats' in the town each weekend. The Springs Branch is a short and extraordinary private canal which runs into a deep ravine at the foot of the castle walls. It was built to provide egress from the stone quarries by the castle. From Skipton to Bingley the canal follows the eastern bank of Airedale, an exhilarating journey along the grassy hillside, with superb views over the dale. The long (seventeen miles) pound from Gargrave ends at Skipton, marking the virtual limit of the pastoral surroundings.

151

BINGLEY TO LEEDS: The 'Bingley Five-Rise' is a staircase of wide locks that is known by name, at least, throughout the canal world. The change of level is indeed dramatic, the canal falling sixty feet in a mere five lock-lengths. But only a few yards on is a further three-step staircase, and from here down to Leeds most of the locks are in similar formations of either three or two. The swing bridges also conspire to retard progress along this length. It is an industrialized stretch, though in the nicest possible way, with the canal getting a good view of the valley's industries and towns without ever being overawed by them. There are some particularly large and elegant stone-built mills at Saltaire, with an adjacent estate village built for the workers, and an unexpected Italianate church rising from its lawn by the canal.

Despite the remorseless urbanization of the valley as it nears the city of Leeds, Airedale seems to cling on to its woods and farmland until the last possible minute—a factor which provides some strange contrasts of scenery. To enter Leeds by canal, which eventually and reluctantly knuckles down to the rigours of the urban environment, is an unbeatable way to arrive. The canal joins the head of the *Aire and Calder Navigation* in the town centre.

BOATING ON THE LEEDS AND LIVERPOOL

The industrial nature of much of the canal's surroundings and the scale and type of the locks have often deterred people who are new to canal boating from starting with the *Leeds and Liverpool Canal*. It is certainly rather a 'heavy-duty' navigation. Most of the locks are deep—anything up to twelve feet—which makes the business of locking through a very different proposition by comparison with the shallow, narrow locks found on many smaller-gauge canals. The paddle gates are bulky affairs with huge, sliding wooden paddles built into them and operated by fixed (built-in) windlasses. The ground paddle gear is usually encased in a wooden framework: the vertical spindle means that it has to be wound open or closed in a horizontal plane, although another type is thrown wide open at a single heave on a wooden bar. Generally, the bottom paddles are kept padlocked as a pre-caution against vandalism: boaters have to procure a special T-shaped key.

On the eastern end of the canal, from Bingley downwards, the staircase locks make things a little more complicated, but it is not difficult to learn the routine for negotiating them efficiently without either flooding the lockside or, at the other extreme, stranding one's boat on the floor of the lock chamber. The swing bridges in this area should also be used with some respect, for some of the minor roads that they carry become well frequented during weekday rush hours.

John Rennie's aqueduct near Lancaster.

One important point about the locks on the *Leeds and Liverpool* is that it is impossible for full-length narrowboats to navigate east of Wigan, or along the Rufford Branch: the locks are only sixty-two feet long. There has been a rapid rise in the number of boating businesses established along the canal in recent years, almost exclusively within Airedale, and there is now a good range of chandlers' shops, repair yards and hire cruiser firms. One has even pioneered a powered pontoon for hire by caravan owners who wish to cruise the canal in their own caravan. It is also possible to go on a day trip boat, the *Apollo*, which operates from the Shipley area.

Lancaster Canal

The link between the *Lancaster Canal* and the *Leeds and Liverpool* is sometimes difficult to comprehend if one looks at a current map of the former canal. Such a map shows clearly that the *Lancaster* terminates on the north side of Preston, eighteen miles north of Wigan. But originally the two canal routes overlapped for a long time, and in theory part of the *Leeds and Liverpool* (from Wigan to Johnsons Hillock) is still part of the *Lancaster.*

The *Lancaster Canal* was planned twenty years after the *Leeds and Liverpool* and was originally designed to run from Kendal, in the north, through

Lancaster and Preston, across the River Ribble, and down to Wigan and beyond. It was intended to carry coal north from the Wigan coalfields in exchange for agricultural produce from the flat farmlands of northern Lancashire and slate from quarries near Kendal.

It was a promising route for a waterway, but the very length of the line and the expense of structures like the great Lune Aqueduct at Lancaster forced the promoters to seek ways of economizing where possible. They arranged for the Leeds and Liverpool company—who were making heavy weather of their connection between Wigan and Burnley—to join the already built length of the Lancaster line at Johnsons Hillock and leave it again at Wigan, ten miles south. This arrangement enabled the Lancaster company to recoup some of their expenditure, but meanwhile they also decided to forego an aqueduct over the Ribble and build instead a temporary tramway between Preston and Walton Summit, north-west of Johnsons Hillock. This naturally involved the inconvenient transhipment of goods at each end, and the company's failure ever to replace the tramway by a continuous waterway link led inevitably to the complete separation of the canal's two sections. Today, much of the link between Preston and Johnsons Hillock has disappeared, so now the waterway stands in isolation. This gives it a pleasant local flavour. In its function too this canal has always been something of a curiosity. The two long level pounds between Preston and Kendal prompted the original company to run an express passenger service between the two towns, with special 'fly' boats drawn by teams of galloping horses. One can still find some of the old stables where the frequent changes of horse teams took place. Another tell-tale characteristic of this canal is the use of stone throughout. The locks and original bridges are almost without exception constructed of hard grey stone which has lasted well—fortunately, for all the structures bear the mark not only of a highly competent engineer but also of a great architect. The style of John Rennie's works on the *Lancaster Canal* are as classic and ageless as so many of his efforts elsewhere in the country.

Another factor contributing to the attractiveness of the waterway is its geographical position, running as it does north–south through northern Lancashire. Although the A6, the electric railway, and now the M6 motorway share this corridor (coming uncomfortably close at times) there are very few cross-routes of any significance. The wildlife population benefits greatly from this lack of interference.

The canal today has been pruned at each end. Its southern terminus is now near a basin at the end of a rather scruffy cutting in Preston, an aqueduct that used to take the canal over a road into the heart of the town having now disappeared along with the last three quarters of a mile of the waterway.

154

Tewitfield Locks—doomed by the construction of the M6 motorway—which still feed water down to the remaining length of the *Lancaster Canal*.

But much more important is what has happened at the northern end, where fourteen and a half miles of the canal are now closed to navigation. It is true that some of this length has been extinct for many years, but since the planning of the M6 motorway, a further drastic shortening has taken place. Thanks to the very poor condition of Tewitfield Locks, the top section of the canal had been only little used for some time, and it was perhaps grimly inevitable that the new motorway should sever it in at least one place. In the event, the canal was cut not once but six times by the M6 or its approach roads. Not only did this chop it into many short sections, making it less attractive to canoeists and others, but the actual road crossings themselves have been executed with no thought at all for the possibility of even walkers wishing to cross. (The only concession to the canal is shown by the culverts that allow essential water supplies to flow down to the lower portion of the canal south of Tewitfield.) And just to add insult to injury, the Kendal Link Road has just been completed, cutting the canal twice more, either side of Hincaster Tunnel.

All this is a shame, because the *Lancaster Canal* becomes prettier and more interesting as it pushes north into Cumbria. North of Carnforth, the extra height of seventy-five feet gained at Tewitfield enables the canal to join the rugged green fells that introduce the start of the Lake District. But the rest of the canal is attractive too, in a different, less conspicuous sort of way. It leaves the outskirts of Preston for the flat farmlands of the Lancashire Plain called the Fylde, a land of—amongst other things—strong smelly cheeses and nine-pip domino games. The hills of the Forest of

Locking up into Glasson Basin.

Bowland stand watch as sturdy stone aqueducts carry the navigation north-wards over the rivers Brock, Calder and Wyre. The canal and the coast converge towards Lancaster, and a flight of six broad locks leads a branch down to Glasson Basin on the Lune estuary. Beyond the dark stone build-ings of Lancaster is a great aqueduct over the River Lune; then the canal almost touches the sea itself at the vast sands of Morecambe Bay. North of Hest Bank, the canal passes Carnforth and winds round past the hamlet of Borwick before coming to a dead stop at the foot of a road embankment beside the M6. Beyond are the forlorn Tewitfield Locks, with the road traffic roaring along just over the hedge. To the north, the disused canal picks its way through the sheep-dotted hills. A smart zig-zag takes it through Hincaster Tunnel beyond which the canal is no more than a long depression along the hillside, holding no water at all. Further on, the canal bed is infilled, and the surviving bridges serve only as a reminder of the waterway.

Lune Aqueduct, Lancaster: John Rennie's magnificent, five-arched masonry structure carries the canal over the River Lune. The high semi-circular arches carrying the classical stone balustrading show a care for architectural design which is evident right along the canal. There is another, modern aqueduct over the A683 at the south end of the Lune Aqueduct.

Lancaster: Near the centre of the town, the former company headquarters are indicated by old canal basins and stone-built canal stables on the opposite bank. Just the other side of some moorings is a ruined canalside building which used to be an important boat-repairing works in the old fly-boat days.

Glasson: A tidal dock and a very large basin at the edge of the wide but shallow Lune estuary, Glasson Dock is busy with sea-going vessels ranging from yachts to large coasters; this is a pleasing and unusual conjunction of canal and sea. The basin is the *Lancaster Canal*'s exit to the sea. It is noticeable that the six locks on the branch have exactly the same pattern of sliding wooden paddles in the gates as those on the *Leeds and Liverpool Canal*.

Garstang: The Lancaster Canal Boat Club is based at this attractive old basin on the edge of a small and pretty town. The old tithe barn has been rescued and turned—with great skill—into a restaurant and museum of

Hincaster Tunnel, on the *Lancaster Canal*.

The wide bridges on the *Lancaster Canal* left plenty of room both for broad-beam boats and for hurrying horses.

agricultural implements. Nearby is a substantial but elegant single-arched aqueduct over the River Wyre.

Hincaster Tunnel: The rich overhanging foliage at each portal contrasts with the impeccable state of repair that this long-disused tunnel still enjoys. One may gauge the quality of the workmanship by the fact that the neatly cut stone facing remains intact and unpatched, while the tunnel itself continues to hold water as it was always designed to do. Another rare feature is the stone rubbing strake along each side to protect the boats. The towpath's separate course over the top of the ridge is kept private by a little 'underpass' at each farm crossing. One feature at Hincaster that might have proved tricky to the navigator is the very sharp bend at the west end, although this would have been eased by the exaggerated width of the canal at this point. Today the tunnel is unfortunately bracketed by the Kendal Link Road, but it is still very easy to get at from either end from local lanes.

Tewitfield Locks: Tewitfield is a good place to go and get indignant about modern progress. Of these eight broad locks, the only ones on the canal's main line, only the deep stone chambers are left now: the gates have given way to concrete weirs, for the locks still form part of the water channel that supplies all of the navigable part of the canal from Killington Reservoir. The locks are cut off from the new navigation terminus by the A6070, but it is possible to walk round and find—usually—a small huddle of boats on the other side. The motorway traffic thunders alongside, oblivious, but there's a pub nearby as if to offer compensation.

BOATING ON THE LANCASTER CANAL

The *Lancaster Canal* is a particularly relaxing canal for a boating holiday. There are only six locks in all, and these are off the main line. There are no tunnels and no swing bridges (the big stone bridges offer a very comfortable headroom of eight feet), and there is never a vast amount of traffic about, if only because the canal is unconnected to the rest of the network. This determines to a certain extent the types of boats that use the canal: any narrowboat would have to be transported overland to reach it, unless it risked a passage along the Lancashire coast from the *Leeds and Liverpool Canal*. Mostly to be found on the *Lancaster Canal* are modern fibre glass cruisers and runabouts, although some of the old broad-beamed barges once standard on the canal may still be seen—including one that runs day trips from Preston. It is predominantly a rural canal for, apart from Preston, the only town of any size is Lancaster—which is more interesting than industrial. There are several boatyards on the canal, principally a group of small ones in the Catforth area (towards Preston), and a marina at Galgate.

8

Canals Around Manchester

ASHTON CANAL

PEAK FOREST CANAL

MACCLESFIELD CANAL

CHESHIRE RING

The 'new' route from Manchester to the Potteries, this line of narrow canals rises just above sea level to over 500 feet, progressing from the darkest corridors and grimmest of industrial valleys to the green, open sides of the Pennines.

Ashton, Peak Forest and Macclesfield Canals

These form the shortest route between the Potteries and Manchester, a route which was only formed by the completion of the *Macclesfield* in 1831, and which for the first time presented the *Trent and Mersey* with substantial competition. The route embraces some of England's finest countryside and some of its most intensely industrial urban areas. It contains the highest length of navigable canal in Britain, and descends almost to sea level in Manchester. It is a route that has been threatened for some time due to the closure of the lower two canals; but now restoration of both has breathed new life into all three waterways.

The oldest of these canals is the *Ashton*. An Act of Parliament in 1792 authorized the construction of a canal from Manchester's Piccadilly to near Ashton-under-Lyne and Oldham; a series of other Acts preceded construction of various lengthy branches. Meanwhile, the *Rochdale* (one of the two now-defunct trans-Pennine routes) continued the *Ashton*'s line down to meet the *Bridgewater Canal*'s eastern terminus in the centre of Manchester. With its numerous branches (the two main ones were to Stockport and Hollinwood), its connection with the *Huddersfield Narrow Canal* (the second of the now-dead trans-Pennine routes), and, in 1800, the arrival of the

Marple Aqueduct.

Peak Forest Canal, the *Ashton* became one of Manchester's busiest waterways. The paired upper locks on the *Ashton* bear witness to the once-heavy com-mercial traffic with which the canal once had to cope.

The *Peak Forest* was built not very long after the *Ashton,* having been authorized in 1794, the biggest boom year for canal schemes. The canal was intended to run from the upper end of the *Ashton Canal* to east of Chapel-en-le-Frith in the Peak Forest. The idea was to tap the great limestone deposits in the region—an ideal source of building materials for Manchester's rapid expansion. The canal's promoters were shrewd enough: those deposits are still being worked today, and show little sign of giving out, but the canal was not taken as far as was originally intended, for it was deemed more prudent to overcome the steep rise in the canal line's last few miles by constructing a light railway or tramway from Bugsworth (now Buxworth) eastwards. This was duly built, and entitled the Peak Forest Tramway. It was a crude yet thoroughly sensible solution, using horses to draw small wagons along this very early form of railway; there was even a steep 500 yard incline, and two tunnels. But a later and more adventurous piece of engineering concerns Whaley Bridge. Here, at the canal's other southern terminus, a connection was planned to join the waterway with the *Cromford Canal,* way over on the other side of the mountainous Peak District. Very wisely, a canal was not attempted: a standard gauge railway was built instead and acted as a direct link between Whaley Bridge and a wharf not far south of Cromford itself. Opened in 1831, this remarkable railway featured numerous inclines, and although ludicrously primitive in con-ventional terms it worked efficiently and extremely usefully until it was finally closed in 1967. It thus comfortably out-lived the Buxworth tramway system and perhaps explains why the Whaley Bridge terminus is regarded nowadays as the canal's main line, Buxworth being considered a subordinate branch.

The *Macclesfield Canal* arrived much later. It was not even seriously discussed before 1824, and even then the promoters were not sure initially whether to make it a canal or to try their luck with one of the new-fangled railways. Fortunately it was built as a canal: authorized in 1826 on a line surveyed by Thomas Telford himself, it was completed in 1831. The *Macclesfield* was thus one of the very last principal canals to be built in this country; and although impressively engineered, its chance of withstanding the onslaught of the railway competition that was just around the corner was prejudiced by its isolated position and the numerous locks that bracketed it on adjoining canals.

Eventually the canal sold out, along with its neighbours on the *Peak Forest* and *Ashton* canals, to the fore-runners of the Great Central Railway. But the canal meanwhile had put the silk and other textile mills of the Macclesfield area in touch with the market in Manchester, as well as opening up an outlet for the coal-producing area near Higher Poynton. The canal continued to be well-used, albeit not very profitably, until the end of the nineteenth century.

But the end of commercial traffic did not help the *Macclesfield Canal's* chances of survival, nor those of the *Ashton* and *Peak Forest*. The latter's locks and the great aqueduct at Marple were a permanent maintenance liability. The *Ashton Canal* too was extremely heavily locked and suffered badly from an unreliable aqueduct in Manchester, so eventually the locks on the *Ashton* and the lower section of the *Peak Forest Canal* (north of Marple Junction) became unofficially disused. Nature and the attention of vandals and sundry rubbish dumpers soon secured the closure of these canals, although canal enthusiasts were determined not to allow the situation to become permanent and mounted a ceaseless campaign. 'Restore the Cheshire Ring!' was the cry. It is worth outlining here the story since then.

Cheshire Ring

The 'Cheshire Ring' is a 100-mile circle of waterways consisting of the *Ashton Canal*, the *Macclesfield Canal*, and parts of the *Bridgewater, Trent and Mersey, Rochdale* and *Peak Forest* canals. It is a circle that has been incomplete since the *Ashton* and *Peak Forest* canals became impassable; and the

A relic of the old days.

neglect of the *Rochdale Canal* locks in Manchester widened the gap. The Inland Waterways Association has fought a long and tireless campaign to get all three canals reopened to navigation. They even went as far as the House of Lords to try and get the British Waterways Board officially taken to task for failing to maintain the navigation, but were foiled by the hurried passing of the Transport Act (1968) which extinguished the public right of navigation on Britain's canals. Undaunted, the IWA mounted massive weekend clearance operations on the *Ashton Canal* in both 1968 and 1971, which showed that canal enthusiasts were willing to come from all over the country to lend a hand at the filthy task of cleaning out a derelict canal. This reminded those who lived in places like Droylsden and Ashton-under-Lyne that even if *they* didn't care for the local canal, others did. Indeed, there was never any overwhelming physical obstacle to restoration, only the difficulty in persuading certain people that the best—and cheapest—solution for a disused canal is usually to restore it to full navigable condition and use it. The campaign eventually succeeded in 1971, with agreement between the British Waterways Board and various local councils that restoration should be carried out. The efforts of the IWA—and of the Peak Forest Canal Society, which over the years had been quietly renewing the worn-out lock gates at Marple—had at last borne fruit, and sure enough the *Ashton* and the *Lower Peak Forest* canals were duly reopened to navigation in 1974. Unfortunately, problems of persuasion still remained in connection with the one mile of the *Rochdale Canal* that is part of the *Cheshire Ring*. This is one of Britain's few privately owned canals, and the Rochdale Canal Company was perhaps a little taken aback to find that its last mile of canal, containing nine locks, was intended to become part of a through cruising route. The company had to find over £60,000 to put the locks back into good condition again. At the time of writing (1976), the *Rochdale Canal* was still under repair. When this short link is reopened, the *Cheshire Ring* will at last be a reality, and the busy waterway from Marple down into Manchester will remain a living reminder of what hard work and tireless campaigning can achieve.

Ashton Canal

The long saga of the *Cheshire Ring* campaign by the IWA thrust the *Ashton Canal* into the limelight in no uncertain manner. Indeed, it is probably true to say that (with the possible exception of the *Kennet and Avon Canal*) there is no other canal in this country which has been the subject of so much heated discussion, particularly in recent years. The IWA's zeal was in no way diminished by the apparent unattractiveness of the *Ashton Canal*. The water-way has always been an unashamedly industrial one, although it is by no means a mere backwater. Much of Manchester's wealth was created in the mills alongside, and the many branches and private arms that used to lead off the *Ashton*'s main line (they are all closed now) made it one of Manchester's principal trade arteries of 150 years ago.

But until recently it certainly had a most unpleasant appearance, partly because of its stagnant, rubbish-filled waters, and several disintegrating structures, and partly because of its sullen, despoiled surroundings. Now things have changed. The canal is once again navigable, and the deep narrow locks are all in good shape. The old bicycle frames and prams and things have been taken from the canal bed and pleasure boats are once more to be seen using the waterway. In short, the *Ashton Canal* has been brought back to life, and its revival can hardly fail to lend weight to other canal restoration campaigns.

164

Restoration has also revealed that the *Ashton Canal* has its charms after all. Some of the bridges spanning the paired locks have an unexpected air of lightness and elegance. Newly painted locks and recently installed canal-side seats and saplings have also helped to cheer up the scene, while the increasing interest shown by canalside factories and works in tidying up 'their' bit of the waterway banks can do nothing but good to the canal environment. But mostly the *Ashton Canal* is characterized by its industrial surroundings, the tall chimneys of the cotton mills and the dark, enclosed corridors of early nineteenth century industry.

SEEING THE ASHTON CANAL

It is difficult to recommend any particular spot on this canal as worthy of special interest: it is only six and a quarter miles long, so perhaps the best thing to do is to walk the whole of its length. There are plenty of good train and bus services for the return trip. Otherwise, have a look at the Manchester end: the most intensely industrial stretch is between Ancoats and Beswick locks, where there is still a moribund branch to be seen (the Islington Branch). Below Ancoats Locks an aqueduct goes over a street, and nearby is the junction with the *Rochdale Canal* which merits a visit, even if the old basin at the junction has given way to a car park.

Another obvious place to see the *Ashton Canal* is Dukinfield Junction, where the *Peak Forest* makes a T-junction at a long roving bridge. Just to the east are the lonely moors of the Pennine Hills. In the foreground is the industrialized Tame valley; the river passes under the *Peak Forest Canal* at the junction and the *Ashton* continues up the valley for a further half mile before joining the *Huddersfield Narrow Canal*. Although the latter is now beyond redemption, with the first of its locks dismantled, water continues to flow down from the hills to feed not only the *Ashton*'s locks but also the British Waterways Board's industrial customers who use the water for cooling purposes.

Peak Forest Canal

Geographically, this is a canal of tremendous contrast, with one end in the grimiest of Manchester's outskirts and the other in the foothills of the Pennines. It is an excellent waterway along which to walk, boat or bicycle, and the recent reopening to navigation of its northern section is a real milestone, not only in the field of waterway restoration but also in terms of what

From Manchester up into the foothills of the Pennines: the *Peak Forest Canal* at New Mills.

An old 'horse tunnel' under the road at Marple. The grooves in the stonework (right) were cut by taut hauling lines over decades of horse power.

the 'Leisure Age' is really about. 'Amenity', 'outdoor leisure pursuits', 'industrial archaeology', 'environmental improvement'—all these abstract concepts come to life with the new *Peak Forest Canal.*

At Dukinfield Junction, the *Peak Forest Canal* leaves the industrial corridor that encloses the *Ashton Canal* and sneaks off through a maze of railway lines to find its own, secretive course to the south. The despoiled valley of the small River Tame gives it an unpromising start, but the water‑way manages to skirt the edge of Hyde's faceless sprawl and regain a secluded and twisting course along a hillside. Then it dives between two isolated mills, through the short Woodley Tunnel and quickly past Romiley before emerging on the steep wooded hillside of the Goyt valley. (The rivers Goyt and Tame combine nearby to form the headwaters of the River Mersey.) Another longer tunnel takes the canal beneath a farm and back onto the isolated hillside, where the narrow trough of Marple Aqueduct carries the navigation high over the Goyt valley before it slips under the adjacent railway to reach the first of sixteen locks up to Marple. At the top of the flight is Marple Junction where the *Macclesfield Canal* joins, and from here the *Peak Forest Canal* follows a very steep hillside all the way to New Mills and Whaley Bridge—a canal of distant mountain views and tunnel‑like avenues of overhanging trees, which ends in a flourish at an historic terminal building in Whaley Bridge. The line to Buxworth diverges just to the north, crossing the valley on a tree‑shrouded aqueduct before reaching the overgrown wharves of Buxworth Basin.

PLACES TO SEE ON THE CANAL

The *Peak Forest Canal* is attractive throughout virtually all of its length south of Hyde, but there are three places of outstanding interest: Marple, Whaley Bridge and Buxworth.

MARPLE: The canal's passage through the suburban township of Marple is remarkable mainly for the harmony which the town itself manages to achieve with a canal whose chief feature is a dense flight of locks complicated by expansive side pounds—in a place where level space is at a premium. Neatly mown private lawns, unfenced, encircle the side pounds, providing a haven for ducks; while adjacent paths and roads, equally open, allow passers‑by the freedom to study closely the locks and their operation. Clearly, today's planners—or inhabitants—have picked up the thread of yesterday's ideas with enthusiasm and conviction. The locks themselves are extremely deep by English standards, rising over 200 feet in only sixteen steps, and this enables them to disappear under road bridges and appear as

Marple Locks, *Peak Forest Canal*.

if from nowhere on the other side. The canal's resourceful builders were clearly up to the mechanical challenges that the steep ascent required, and clever devices to substitute for gate balance beams, as well as one miniature tunnel for the towing horses, have been used here, while the wide side pounds themselves are a way of enabling as steep a flight as possible to be built, providing at the same time a reasonably flexible water supply for the boat traffic. Individual points of interest are the great aqueduct at the foot of the locks, the skewed railway tunnel that slants underneath the flight, the magnificent canal buildings further up, and the environs of the junction at the very top—with a good turnover bridge, an old maintenance yard and plenty of boats.

WHALEY BRIDGE AND BUXWORTH: These two sites of interest were, with Marple, the chief focus of activity on the *Peak Forest Canal* in the old days, with large quantities of stone being unloaded from railway wagons into boats at the respective wharves. Whaley Bridge is the smaller terminus, but most likely it was the more sophisticated. There is a fine stone building straddling the water at the very top dated 1832 at one end and 1910 the other. The water bubbling up at the older (southern) end is the canal's principal feeder from Combs and Toddbrook Reservoirs, just up the valley. It is easy to trace the course of the High Peak Railway, part of whose course through Whaley Bridge is now a public footpath. The first of the many inclines begins just beyond an iron bridge over the River Goyt;

an old points lever is still fixed in the ground, and in the adjacent works can clearly be seen the actual rails of a private siding.

Buxworth Basin is nearby. The easiest thing is to walk along the canal from Whaley Bridge, past the junction, over the swing bridge and back down to cross the valley on the aqueduct. The sweat of volunteer labour has cleared all of this length past the canal buildings, to just beyond the gauging lock (a level lock designed to facilitate the calculation of tolls). But beyond the new terminus, all of the Buxworth Basin complex is a scene of desolation, the basins overgrown with long-established trees and weeds which thrive on the permanently wet ground. The tramway's path up into the hills can easily be traced by those with time to spare. There is a minor road back over the hill to Whaley Bridge which completes the triangular two and a half mile walk. The quickest way to regain the canal terminus itself is by walking down the inclined plane through the town.

BOATING ON THE PEAK FOREST CANAL

Navigating the *Peak Forest Canal* can be an exhilarating experience, for the upper level of the waterway is the highest pound on the navigable network—over 500 feet above sea level. It is also in a side cut, which gives it good views across the Goyt valley and privacy from its surroundings—which is useful, for although the main A6 road runs very close it does not intrude at all. Working through the restored Marple Locks is also something of a novel experience after their years of disuse. The locks themselves are very deep and it would be more than a little complicated to work through them alone if you had to shut the gates after you. But the locks are very conveniently close together, and the rest of the canal is lock-free. Other relevant points are perhaps that the several swing bridges towards the southern end of the canal can be very stiff, and at present there is rather a dearth of boating facilities on the canal. The top of Marple Locks is probably the best place from this point of view. The lower section of the canal is fairly straightforward and the two tunnels give no particular difficulty.

Macclesfield Canal

The official start of the *Macclesfield Canal* is not, as one might expect from a map, at the canal junction just north of Harecastle Tunnel, but at a point a mile to the north, at Hall Green (a junction insisted on by the Trent and

169

Mersey Company, who were so jealous at their water supply being lost to the much newer canal that they demanded two separate stop locks, and took the trouble of building the canal from Hardings Wood to Hall Green, incorporating a curious flyover). From here, the *Macclesfield Canal* follows an isolated course through a superb landscape of steep green hills on one side and distant flat farmlands on the other. The hills are the western edge of the Pennine Hills and the farmlands those of the Cheshire Plain. All of three main towns along the way—Congleton, Macclesfield and Bollington—are well-established textile milling towns, mostly attractively built in stone and all on a level well below that of the canal which passes them effortlessly on aqueducts or embankments. The canal's relatively late construction date shows itself in the other substantial earthworks that keep the canal on course despite topographical irregularities, and in the locks that are neatly grouped into a single flight at Bosley, near one of the two reservoirs that feed the canal. The locks put the canal on a contour almost 520 feet above sea level—making it the highest navigable canal level in Britain. North of Macclesfield, the canal is joined by an abandoned railway, and once north of Bollington it enters a minor, worked-out coal-mining district; the short branch at Poynton used to serve a colliery. Subsidence from the old workings has meant building up the canal banks over the years; so the water is deep and the bridges not what they were (having been raised several times to maintain the headroom). At High Lane there is a short arm; and at Marple, the *Macclesfield Canal* meets the *Peak Forest* as both prepare for the long descent into Manchester.

One of the chief glories of the *Macclesfield Canal,* apart from the scenery, is the architecture of the structures and the quality of the materials used. The bridges are mostly uniform flattened arches in warm brown stone; but there is also a sprinkling of light swing bridges. The best are the excellent roving bridges—with the towing path curving evenly round the parapet and back down to the water's edge. Such bridges—and the locks—would have made for a smooth and steady progress along the canal in the days of horse-drawn boats.

PLACES TO SEE THE MACCLESFIELD CANAL

The canal runs through beautiful and quiet upland countryside, which is interrupted only by one or other of the four towns between Marple and Kidsgrove. But most of them lie well below the canal level and hardly impinge, except with their tall mill chimneys. There are no true villages as such, although the canal is well served by pubs, particularly at the south end. It is a good canal for a well-timed walk on a fine day.

On the *Macclesfield Canal*.

The locks at Bosley are charming and secluded, a flight built of stone. The locks are believed to be unique among narrow canals in having double gates not only on the bottom but also on the top gates. (Elsewhere on the canal system, double bottom gates and single top gates are the norm, with singles on both where the lock is very shallow.) These locks are deep ones: their twelve steps raise the canal level by 114 feet.

Other good places to see the canal are the main wharf at Macclesfield, normally crowded with boats and overlooked by a massive chimney. In Bollington there are sturdy aqueducts to carry the canal over the town, and there's another, smaller but more attractive one in Congleton. Nearby is one of the best of the canal's roving bridges (the other is at Marple Junction, which is of course full of good things to see: these are noted elsewhere).

BOATING ON THE MACCLESFIELD CANAL

This is a quiet, beautiful canal ideal for pleasure boating and quite un-sullied by industry. It is exceptionally easy for boating as there are only twelve locks and a mere handful of swing bridges. There have never been many boating concerns based on the waterway because closure of the *Lower Peak Forest* has made the *Macclesfield* a cul-de-sac for many years; but now that the *Cheshire Ring* is open again, traffic will doubtless increase on the *Macclesfield* and more boatyards will surely open. At the moment, Maccles-field seems to be the main centre for boating services.

9

Shropshire Union Network

BIRMINGHAM AND LIVERPOOL JUNCTION CANAL

SHREWSBURY CANAL

ELLESMERE CANAL

LLANGOLLEN CANAL

A large portion of the *Shropshire Union Network* has been officially abandoned since 1944, but what remains navigable includes some of the best cruising country in England. Two of the principal limbs were closely connected with the engineer Thomas Telford (they are also intersected by his London–Holyhead road, the present A5), but their route and style of building reflect very different periods in the age of canal construction.

Shropshire Union Canal

The *Shropshire Union,* like the *Grand Union,* is a group of neighbouring waterways brought together under 'one roof'. The *Grand Union* amalgamation was very beneficial to the component canals within the group, but prospects for the prosperity of the *Shropshire Union* members were rather undermined from the very start. The full title of the company, formed in 1846, was the 'Shropshire Union Railways and Canal Company', and the original intention was to convert at least some of the canals into railway lines.

Fortunately, the railway interests never got the upper hand, and many of the company's canals prospered right through to the twentieth century. But decline was inevitable, and the end for much of the group was finally sealed by the passing of the LMS (Canals) Act in 1944. Axed under this measure was the heavily locked Newport Branch from the *Birmingham and Liverpool Junction Canal* down to Newport, as well as the *Shrewsbury Canal* and the complicated system of interlocking canals in eastern Shropshire that emanated 'inland' from the River Severn around Coalbrookdale and Coalport. These early canals were developed to serve local collieries and ironworks.

A stout reminder.

They featured what are now regarded as historic works of canal engineering, including inclined planes, a 'tub-boat' canal and a small iron aqueduct at Longdon-upon-Tern which was probably the prototype for Telford's Pontcysyllte Aqueduct. This virtually self-contained canal system had been contracting for the the best part of a century, and the 1944 closure of the surviving limbs can hardly have come as much of a surprise. But it still seems ironic that the old canals at places like Ketley, Trench and Oaken-gates are fast disappearing under a new town that has been named after Telford himself.

Also closed under the 1944 Act was the entire line from Hurleston Junction to Llangollen, and the length from the Llangollen branch south-west to the River Severn at Newtown. Fortunately, the Llangollen route was reprieved, but the line to Newtown did not escape. Indeed, the 'Montgomery Branch' or as it is now generally known the 'Montgomery Canal' (it was

173

actually composed of the west end of the *Ellesmere Canal* and the whole of the *Montgomeryshire Canal*) had already been closed as a through route by a breach in 1936. The breach was never repaired; the closure was formalized in 1944 and the canal has been decaying peacefully ever since. It is a sad loss, for its thirty-five miles would have made a superb cruising waterway, passing as it does through some of the best Border country to be found; and its loss is made all the more acute by its adjacence to the now-bustling *Llangollen Canal* which it used to join at Frankton Junction. But the *Montgomery Canal* is not without its supporters who refuse to accept that the canal is finished as a navigation. The Shropshire Union Canal Society, which lost the battle to save the Newport Branch, is determined not to let the *Montgomery Canal* go the same way. The Society campaigns for the restoration of the canal, and its members have already achieved the restora-tion of Welshpool Lock and enough waterway on either side to be able to run trips in a passenger boat in summer. Influential support has been en-listed in the campaign, and although there are some formidable engineering obstacles to be faced, there is no doubt that the future possible restoration of at least part of the canal is now on the cards.

THE NAVIGABLE COURSE OF THE SHROPSHIRE UNION CANAL

The navigable extent of the *Shropshire Union* today encompasses the 'main line' from Autherley Junction on the *Staffordshire and Worcestershire Canal* (near Wolverhampton) down to Nantwich, Chester and right through to Ellesmere Port on the *Manchester Ship Canal.* There is also the Middlewich Branch and the line to Llangollen.

THE MAIN LINE: The length from Autherley to Nantwich was originally called the *Birmingham and Liverpool Junction Canal* and was built as a link waterway, a direct route from the Black Country to Merseyside. Built at the very end of Britain's canal age (the canal was not opened throughout until 1835), the canal was the last to be built by Thomas Telford and is a very suitable monument to him.

The waterway runs through rural Staffordshire and Shropshire into the flat dairy farmland of Cheshire before reaching the old terminus of the *Chester Canal* at Nantwich. It is quiet, unspoilt countryside, untouched by railway or busy road, and interspersed with handsome old 'black-and-white' small towns and villages. The canal has a well-founded reputation as an ideal pleasure-cruising route, and forms with the *Trent and Mersey,* the *Staffordshire and Worcestershire,* and the Middlewich Branch a good week-long cruising circuit.

A low-slung aqueduct on the *Montgomery* Branch.

The canal manages a very straight course through the undulating land-scape—although this is never so straight as to become monotonous to the traveller. The directness of the route is achieved by the canal cutting ruth-lessly through the hills, and striding across the valleys on substantial em-bankments that pay little regard to the natural lie of the land. In this respect, the canal has more in common with the railways than with its own pre-decessors. The cuttings are very impressive indeed—long, narrow and deep. Their sides are often almost sheer, and festooned with trees and other vegetation that somehow manages to subsist, making the cuttings dark, dank and a little eerie. Where roads cross, the traffic is carried unseen by tall bridges high above the water. The canal is shut away in such cuttings for up to a mile or more at a time; its re-emergence into open country, with the distant views of The Wrekin over to the west, is something of a relief.

The *Birmingham and Liverpool Junction* is divided into long pounds by locks which, except for the stop lock at Autherley and an isolated one at Wheaton Aston, are grouped into flights. There are five at Tyrley, five at Adderley, fifteen at Audlem, and a final pair at Hack Green. All are standard narrow locks—although this is a relatively 'modern' canal, it was clearly decided that the reduced construction costs, reduced water require-ments and increased speed of passage for single boats (in the days before motor-and-butty pairs) more than justified the narrow locks. The Tyrley locks are the most attractively situated, the bottom locks being almost roofed over by a tunnel of trees, and the top lock framed by a hump-backed bridge and overlooked by a little row of canal cottages. This waterway is particu-larly rich in minor canal buildings at wharves, locks and maintenance

175

Telford's iron aqueduct over the London–Holyhead road. The *Birmingham and Liverpool Canal* later became the trunk of the *Shropshire Union Canal* system.

Tyrley Top Lock, *Shropshire Union Canal.*

yards, not to mention the solitary canal pubs that are still to be found here and there.

North of Audlem Locks, the countryside flattens out into the Cheshire Plain—rather desolate scenery. The canal terminates near an iron aqueduct at a basin just outside Nantwich. The old basin with its one-time cheese warehouse is now a crowded base for hire cruisers.

From Nantwich to Chester the canal is the much older *Chester Canal,* a wide-locked waterway built to give the local salt and cheese-producing area of southern Cheshire an outlet to the tidal River Dee at Chester. So the canal drops down gradually through low-lying farmland. There are many boat-yards along this route, trading on the large variety of canal routes open to someone starting from this area. Once in Chester, the canal slices through the town in a steep, deep cutting and then descends a rare staircase of three locks carved out of the solid rock. There is a short and rather tired-looking branch into the River Dee nearby, but the main line of the canal continues for another five level miles across the head of the Wirral to Merseyside. This northern pound was once a detached part of the *Ellesmere Canal* (see below), but the idea of the small country town of Ellesmere being in any way related to Ellesmere Port is today almost unimaginable. The latter is a world of oil refineries, motorways and sundry large industries; the terminus of the former *Ellesmere Canal* has been devastated by time. Until recently there was a famous range of Telford-designed warehouses, but the best of these have been destroyed by fire, and the area is more than a little decayed. But the gloomy scene is enlivened by the inescapable presence of the *Manchester Ship Canal.* It is all a far cry from the other end of the *Shropshire Union* at Autherley Junction.

Llangollen Canal

This is no place to try and unravel the complicated historical background of what today forms the *Llangollen Canal:* suffice it to say that the present route is very much an amalgamation of various canals and branches which were never originally intended to form this line. The layout of the various junc-tions along the way give few clues as to which canal came first.

Today the route is officially the Llangollen Branch of the *Shropshire Union Canal,* and indeed it is, or was, a subsidiary branch of the Shropshire Union system; but the decline of its neighbours, the length of the waterway (forty-six miles) and its fame as an ideal introduction to the joys of canal

Gnosall: the *Shropshire Union*'s main line has its fair share of waterside pubs.

On the *Llangollen* Branch. The bascule-type bridges on this canal are much easier for road traffic than the blind hump-backed bridges found elsewhere in the country.

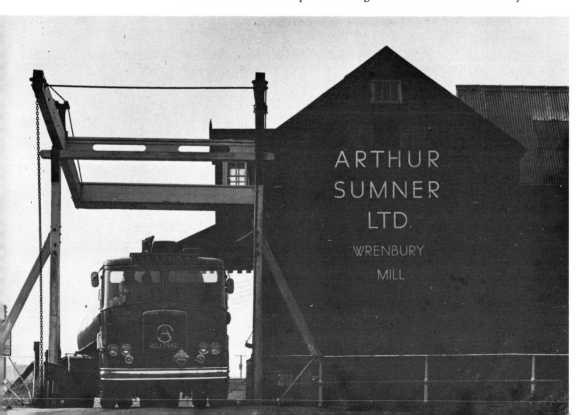

boating have combined to modify history and give it a separate title. Some call it, rather confusingly, the 'Welsh Canal'; but more usually it is known as the 'Llangollen Canal'.

The canal is lucky to survive. It was legally abandoned in 1944, along with so much of the Shropshire Union system, but because of its long-standing usefulness as a water supply channel right the way down from Llantisilio to Hurleston, the entire line was saved. Since then it has become Britain's best-known narrow canal.

The *Llangollen* is rightly famous. Its course stretches from the soft green farmlands of Cheshire through the intimate countryside of northern Shropshire and up into the Welsh mountains, taking in areas as diverse—and close—as Whixall Moss and the little tree-fringed lakes that lurk in the hills around Ellesmere. There are no towns along the way and hardly a whiff of industry. The canal is strictly a tiny rural byway, creeping past the quiet villages and their black-and-white houses with the utmost discretion—except, that is, for the bridges, many of which are immodest but endearing wooden bascule bridges with overhead balance beams. Most of the locks are located at the east end of the canal, including a compact flight of six at Grindley Brook. There are a lot of farms along the canal too, many of which 'cash in' on the boating boom by selling their home produce to the passing trade. There are two branches along the *Llangollen*—one from the lonely junction on the embankment at Whixall Moss runs for a mile or so to a former clay pit, now a canal boat marina. This mile is all that remains navigable of the old Prees Branch. The canal's other branch is the short arm to the town wharf at Ellesmere.

Further west, the canal changes dramatically as it meets the east edge of the Welsh hills. At the border village of Chirk, a narrow trough leads the canal over a great masonry aqueduct seventy feet high over the river Ceiriog. This is followed immediately by the first of two tunnels. The Pontcysyllte Aqueduct is a little further, and crosses the River Dee at the startling height of 120 feet. A now-truncated branch at the north end of the aqueduct was originally intended to continue the *Ellesmere Canal* to Chester and the Dee, but the intervening canal was never built, even if the Chester–Ellesmere Port section was. However, the company's feeder channel from Trevor along the mountainside to tap the Dee's waters was cut and a weir constructed (the Horseshoe Falls) to provide a constant supply of water. Happily, this was made as a navigable feeder, thus giving canal access to the town of Llangollen. It is a spectacular length of canal, following a very precarious course along the wooded mountainside. In places, the waterway is rendered so narrow by the steepness of the slope that the channel is not even wide enough for two boats to pass. But the water, fresh from upland Wales, is clear as a bell.

179

Pontcysyllte Aqueduct—completed in 1805 and still going strong. The dovetailed iron joints helped to prevent leakage. The success of Telford's design for an iron bridge resting on stone columns was later of great use to the railway engineers.

PONTCYSYLLTE AQUEDUCT: This most famous and perhaps most astonishing of all Britain's canal aqueducts was opened in 1805, and is one of Telford's greatest feats of engineering. The enormous length (1007 feet) and height were achieved by the entirely new technique of using a water-tight iron trough to contain the canal and placing this along the top of a series of masonry pillars. This departure from the earlier 'non-iron' approach had great advantages in the amount of support necessary for the actual water-way channel, enabling it to be carried at a great height over the valley on the same level as the mountain-hugging canal on either side. A further refine-ment was to cantilever the towpath over the canal, thus providing as large as possible a cross-section of water in the trough to assist the passage of boats. These features make Pontcysyllte not only the biggest aqueduct in this country but also one of the most elegant.

THE MIDDLEWICH BRANCH: The *Chester Canal* was initially designed to run from Chester to Middlewich, but in the event it was built to Nant-wich instead, and a fresh link to the *Trent and Mersey Canal* at Middlewich was not made until the early 1830s. It is not of especial interest, being a quiet and remote waterway in sparsely populated farmland; there are several extremely deep locks along the route. Near the small village of Church Minshull is a small aqueduct over the River Weaver: through the trees from this point can be seen Winsford Top Flash, the limit of navigation on the Weaver.

IO

The Severn Basin

RIVER SEVERN

GLOUCESTER AND SHARPNESS CANAL

RIVER AVON

STRATFORD-UPON-AVON CANAL

WORCESTER AND BIRMINGHAM CANAL

MONMOUTHSHIRE AND BRECON CANAL

Navigable waterways that focus on the Severn estuary: two narrow and heavily-locked canals, the newly restored River Avon, and the great Severn itself. The *Gloucester and Sharpness Canal* forms a bypass to the worst of the Severn's meanderings below Gloucester. The *Monmouthshire and Brecon Canal* is appended to this group.

River Severn

In view of the river's size and obvious regional importance as a great water highway from the Bristol Channel into the West Midlands and Wales, it is perhaps odd that the Severn was for so long an unimproved navigation, upon which boats traded almost entirely at the mercy of tides, wind and rain. The first locks were not installed on the river until the middle of the nine-teenth century, well after the *Gloucester and Berkeley Canal* had been built to bypass the lowest reaches of the tidal river where it becomes an estuary. However, the river's lack of navigation works has never prevented it being used by fairly large boats, many of which were either bow-hauled or sailed, even if their progress along the river was erratic and often delayed for long periods. Abraham Darby's iron smelting works and the whole early in-dustrial area around Coalbrookdale (which is way above the present upper limit of navigation) was entirely based on a navigable River Severn, with large quantities of iron ore being taken upstream in boats from Bristol and Gloucester. Completion of the *Shropshire Canal* in 1792 put the river more closely in touch with the now thriving coal mines and iron industries of that county, and all this added greatly to the trade on the river.

Early legislation was more concerned with building a towing path than with the somewhat formidable task of building locks and weirs. But at long last, in 1842, the Severn Commission was set up and four locks were duly constructed. The lock at Tewkesbury was added later; and the navigation has not changed much since. There has been little improvement of the waterway beyond enlargement of the lower locks. Traffic on the river has declined in recent years, partly due to lack of official enthusiasm and partly because of inevitable changes in the pattern of trade. There was an upsurge in traffic from the 1930s onwards, mainly in oil imported from the ports at Bristol and Avonmouth, up to Worcester and Stourport (not counting traffic to Gloucester, served only by the *Ship Canal*), but this has dropped right away in recent years. Nowadays, virtually the only commercial traffic above Gloucester is a small regular cargo of grain to Healing's Mill on the riverside at Tewkesbury, up the Avon backwater. Otherwise, the Severn is given over entirely to sailing boats, motor cruisers and rowing boats of one sort or another. Yet, for all the pleasure boats that use it, the river seems curiously lifeless without a leavening of commercial traffic, and it is unlikely that this traffic will return.

THE COURSE OF THE SEVERN

The Severn rises high in the Plynlimon Mountains, just under fourteen miles from the Cambrian Coast. It tumbles down through the pines of Hafren Forest, losing height rapidly as it heads eastwards towards England. By the time the Severn reaches Newtown, where it meets the Montgomery Branch of the *Shropshire Union Canal,* it is big enough to have carved a well-defined valley for itself, and the river is down to only 350 feet above sea level although the river bed is still a narrow one and the current flows fast. The Severn runs north-east along its beautiful valley still overlooked by the steep green hills of Wales until, beyond Welshpool, it curves right round to head eastwards to Shrewsbury, becoming a waterway of extravagant curves but still of no great width.

Such are the upper reaches of the Severn that have hardly ever been navigable in the sense of allowing trading boats to work regularly along the river without endless delays due to shallows or other obstruction. From Shrewsbury downwards, the Severn has seen boats since the earliest days of navigation, informal and haphazard though the river may have been. South-east of Shrewsbury, the river wanders broadly through a shallow valley until it reaches the incongruous power station of Buildwas, between the historic river bridge and the old-established industrial settlement on the bank at Coalbrookdale. Steep, wooded hills enclose the river tightly from here down to Bridgnorth, which was once the chief port of embarkation for goods from Birmingham destined for shipment right down to Gloucester and Bristol and thus by sea to anywhere in the world. The river continues south through its narrow, secluded valley down into Worcestershire. Bewdley is another elegant town whose eighteenth century prosperity owed much to its role as a busy river port.

The river starts to be navigable a short distance before Bewdley and about three miles upstream of Stourport. From Stourport—where the *Staffordshire and Worcestershire Canal* enters the river—downwards, the Severn is extensively used by pleasure craft of all types and presents in every way a different scene. From here, the river is increasingly broad and crossed by few bridges. Below the canal entrance is another riverside power station, and some old oil wharves which used to be fed by tanker barges until only a few years ago. On the west bank of the river are some red sandstone rocks with caves dug out of them, and nearby is Lincomb Lock, the northernmost of the Severn Locks, overlooked by a steep rocky hillside. The reach down to Holt Lock and Worcester is a pleasantly wooded one, a broad waterway within a warm and sheltered landscape that displays all the softness of Worcestershire. Here and there are small riverside settlements and the occasional pub, half-hidden in the trees on the river bank. There are locks near Holt Bridge and at Bevere Island. Holt Bridge itself, like most of the bridges on the navigable part of the Severn, is tiresomely narrow to those who cross it by road, but large and handsome from the water.

The river's approach to Worcester runs past the race course; the river bisects the town, and the famous cathedral towers over the water. At the south end of Worcester is the entrance to the *Worcester and Birmingham Canal,* a dumpy little entrance bridge at the foot of the two locks contrasting unfavourably with the graceful lines of the river bridges. Just downstream there is a large pair of locks at Diglis, but the adjacent dock that used to draw so much trade up from the river is now silent and disused—no trading boats reach this far any longer. South of Diglis, the course of the Severn is pleasant enough but there are few points of reference, apart from topographical irregularities like the intermittent woods, occasional steep cliffs rising from the water's edge, and views of the distant Malvern Hills to the west. Upton-upon-Severn is a refreshing interruption, an attractive small town with its face very much to the river. The long wharf is used for pleasure now; once it was there to load and discharge Severn Trows. A couple of miles below Upton Bridge, the M50 motorway crosses on a large viaduct; the next bridge is just before Tewkesbury and precedes an important junction, for the River Avon turns in to the Severn here. Tewkesbury is found to the east, just across a great meadow that was the site of an important battle in

Gloucester Docks.

1471 during the Wars of the Roses. Tewkesbury Lock is the last navigable lock on the Severn and leads into a semi-tidal stretch (tidal on spring tides only, when the floodwater overruns a weir in Gloucester). Along this reach are a bridge and several riverside pubs, most of which are difficult of access from the river. Eventually the river divides, just north of Gloucester, the wider channel turning west to the now-disused Maisemore Lock, while the navigable channel runs down past piled banks into Gloucester. The entrance lock up into the *Gloucester and Sharpness Canal* is in the middle of the town, and just downstream is the derelict Llanthony Lock. Both the Maisemore and Llanthony branches of the river combine to form a single tidal and unnavigable waterway which is shallow, broad and winding. It is along this part of the river that the Severn Bore is seen, and it was this dangerous patch of waterway that the *Gloucester and Berkeley Ship Canal* (now called the *Gloucester and Sharpness Canal*), was built to avoid.

The River Severn near Tewkesbury where it is spanned by the graceful Mythe bridge, another example of Thomas Telford's work.

BOATING ON THE RIVER SEVERN

The Severn is better suited for sailing and rowing than for motor cruising. Its long, wide lockless reaches are excellent for dinghy sailors and rowers, while anglers find it excellent value. But the very scale of the river and its remoteness from industry and other artificial features tend to make it dull after a few hours of non-stop cruising. The high banks prohibit much in the way of views, while the large stones that often line the banks make mooring a major problem—even at some of the riverside pubs.

In other words, for the average inland waterway man the river is more useful as a way of getting from one place to another fairly fast than for navigating for its own sake. In this respect it is quite the reverse of the Warwickshire Avon. However, the Severn is by no means an ugly river— the few locks are generously decorated with flowers and the few towns along it—Stourport, Worcester, Upton-on-Severn, Tewkesbury and Gloucester— are all of great interest; the river is worth navigating if only to visit some of these places. All offer plenty of moorings, and with each one of them the approach by boat is better than any approach by car.

Gloucester and Sharpness Canal

Built in the early nineteenth century as a bypass to the worst of the River Severn's twisting shallow course, the *Gloucester and Sharpness Canal* is a waterway of noble dimensions, still carrying a considerable amount of freight in coastal oil tankers and barges. Because of the canal's continuing useful existence, Gloucester remains an inland port—which adds an interesting extra dimension to a town already well endowed with things worth seeing. (When it was opened, this canal was the broadest and deepest in the country, and one can imagine the awe in which it was held by contemporary civil engineers.)

The canal had a very protracted start. The promoters had such a struggle to pay for the work required to construct a canal on such an enormous scale that we are lucky the work was ever finished. In the event, it was help from the Government that ensured the completion of the waterway over thirty years after work had started. It was originally intended to rejoin the sea at Berkeley Pill, but it was later decided to terminate it a little nearer Gloucester. Hence the creation of Sharpness, that lonely little port still thriving today as much as ever.

The canal is—perhaps surprisingly for such an important waterway—imbued with considerable charm. Its passage through the flat farmlands of the Severn Valley is often slightly embanked, giving views of the hills that rise from the far bank of the River Severn. The bridges—necessarily of an opening type to cater for the masts of the shipping for which the waterway was designed—are little swing bridges, and the bridge keepers live in tiny but elegant lodges which sport classical pediments and columns; these help to give the canal a very stylish uniformity.

By acting as a navigable alternative waterway to the tidal Severn, the *Gloucester and Sharpness Canal* creates a kind of 'island' comprising mainly salt marshes and water meadows. The absence of any road bridges over the river between Gloucester and the Severn Bridge towards Bristol means that the only traffic crossing the canal is the traffic bound to or from the various small villages within the 'island'. This helps to keep the canal quietly isolated and free from any new road bridges. (It is no accident that the Slimbridge Wildfowl Trust was established here.) In this respect the waterway is as unchanged in its appearance as any that one can imagine. Perhaps the oddest thing about the canal (to anyone used to smaller canals), is the relative proportion of its surface width to its depth. For here is a waterway only twice the width of an average small English inland canal, yet which carries vessels displacing up to 700 tons. Most of the traffic is 'small' seagoing oil tankers which unload at Quedgeley, a little short of Gloucester, but other traffic includes barges which bring timber right up to the wharves at Gloucester Docks, having transhipped at Avonmouth Docks way down river.

The canal's course, roughly north-east–south-west, is not a regular line curving through the landscape, but more a series of straight lines broken up by fairly sharp corners. At one point on the outskirts of Gloucester itself there is a sharp double bend through a cutting; the docks are on the city's

The *Gloucester and Sharpness Ship Canal*—not very wide, but deep enough to get sizeable vessels up as far as Gloucester.

west side and the canal is protected from the River Severn by a lock up from it. From here, it runs fairly straight for two miles, past wharves and depots both used and disused; the double bend is very close (though unseen) to the River Severn. The oil terminal is next to be seen, and then a 'cross-roads' with the old *Stroudwater Canal*. Frampton-on-Severn is marked by the church beside the canal, quite near the Slimbridge Wildfowl Trust which lies between the ship canal and the estuary. The last village is Purton, where there are some timber ponds used formerly for storing uncut timber afloat, before it was towed upstream to the sawmills in Gloucester. From this point onwards, the canal runs right along the river bank of the wide estuary, passing the remains of the old iron girder railway bridge that used to cross the river, and at Sharpness it becomes a dock complex, with a very large lock at the south end into the Severn estuary.

PLACES TO SEE THE CANAL

GLOUCESTER: Gloucester Docks are of interest not only for the colourful sight presented by the variety of boats within but also for the very fine, early nineteenth century warehouses ranged around the principal basin.

At Saul, the *Gloucester and Sharpness Canal* is crossed by the now-defunct *Stroudwater Canal*, the western extension of the *Thames and Severn*. The gate paddles of this lock on the *Stroudwater* are almost identical to the sliding gate paddles on the *Leeds and Liverpool Canal*.

They are massive, perfectly plain and yet a joy to the eye. The docks themselves are less busy than they were, although the increasing size of vessels tends to offset the declining number of craft movements. By the lock, water pours into the canal: it is pumped up from the Severn, for the canal is higher than the river and has therefore to be constantly supplied to compensate for the wastage at each end.

SAUL: Saul is a village at the level junction of the *Gloucester and Sharpness* and *Stroudwater* canals. The latter used to be part of a route joining the rivers Thames and Severn. The *Thames and Severn Canal,* which was abandoned in 1927, ran from Inglesham at the head of the Thames navigation westwards past Cirencester through the two and a quarter mile Sapperton Tunnel and down the Frome valley to Stroud, a distance of just under twenty-nine miles. The *Stroudwater Canal* provided the eight mile link down to the Severn estuary, crossing the ship canal a mile short of the river. There is a disused lock at the junction. (Note the 'horizontal' pivoting paddle gearing on the lower gates, an unusual sight in this area: such paddles are well known on the *Leeds and Liverpool Canal.*)

SHARPNESS: An isolated but fascinating place, Sharpness lives only for the bustling docks; a large tonnage of goods is unloaded here out of ships that displace up to 5000 tons Their passage up the estuary is strictly regulated by tides, for at low water an alarming expanse of sandbanks appears in the estuary. Goods come mainly from Europe, and include a regular timber traffic from Russia and Scandinavia. The Severn is a difficult river navigation at Sharpness, for the current is very fast through this bottleneck in the river's course, and the tidal range is said to be greater than any other place in Britain, while the valley is often blasted by very strong winds from the Bristol Channel. It is hardly surprising that there has been talk of building some form of tidal barrage across the Severn at this point to harness the river's energy. The oldest part of the Sharpness Docks is to the north, with an arm down to the original basin and a disused lock down into the river. The lock house here has a wonderful view up and down the Severn, but it is extremely exposed to the elements and no place to be on a stormy night.

BOATING ON THE SHARPNESS CANAL

It is most important on this waterway for pleasure boats to beware of the commercial traffic that uses it. The ships do not in fact thunder along as they would on a river, but their draft of up to ten feet sometimes causes a certain amount of turbulence, and it is advisable to hold to the bank as they pass.

Since the bridges are all manned swing bridges, it is useful for the staff who man them to know you are on the way. The best method to ensure this is to report to the British Waterways Office at Gloucester or Sharpness before setting out. Weekend opening is restricted: check the times before reaching the canal. Boating facilities are restricted to moorings at Sharpness and Gloucester, and a good chandler's shop on the old arm at Sharpness.

River Avon

The Warwickshire Avon is one of the Midlands' lovelier rivers, smaller than the Severn or the Trent, more intimate than the Thames. It is a river that runs through one of the softest, most fertile valleys of England, the Vale of Evesham, and from its banks stretch fields of not just corn but extensive fruit orchards—strawberries too. It is a tame and gentle river, rarely given to serious flooding. So the banks are low, and the villages that border the river show a certain confidence in their proximity by stretching down to the very edge of the water. The pleasure boats which throng the waterway display a cheerful lack of concern for the elements—something which boaters on the Severn can never afford to do.

The river rises in the Northamptonshire Heights, on the border of that county with Leicestershire and close to the *Grand Union Canal*'s Leicester Section. From there it feeds a reservoir at Stanford and flows south-west, under the *Oxford Canal* at Rugby first, and then the *Grand Union*'s main line between Warwick and Leamington Spa, gathering up the waters of the Leam, the Sowe and the Swift. Three miles above Stratford is Alveston Weir, right across the river, and below it the Avon is navigable right down to Tewkesbury where it joins the Severn.

THE NAVIGABLE COURSE OF THE AVON

From Stratford to Evesham the river is called the Upper Avon; from Evesham Lock downwards it is the Lower Avon and is administered by a different Trust. The three mile reach above Stratford from Alveston Weir to Stratford is not formally part of the navigation.

The Avon in Stratford forms a very important part of the townscape, flowing right past the central parkland that adjoins the Memorial Theatre.

189

In summer, its broad, limpid waters are often crowded with punts and rowing and trip boats—although now that the river has been restored to navigation throughout, motor boats also play an increasing part in the scene. Opposite the tall spire of the church where William Shakespeare is buried is the top lock on the new Upper Avon, a steel-framed structure. Below Stratford the navigation shows signs of its youth, with still fresh-looking banks tidily graded and as yet relatively unlined by the holes of water rats, voles and the like. The steel and concrete sides of the locks too are still refreshingly new. There are locks at frequent intervals, and although the fall of each one is often very small, all were painfully necessary to the reconstruction of the navigation. Many of these locks were built where none existed before—a measure of the enormous scale of the problem that faced the Upper Avon's restoration team. One lock had to be built where the river passes under Marlcliff, a steep rock face. Massive blasting operations on the river bed were required to give the necessary depth.

The countryside that forms the Upper Avon valley is green and quiet, a pastoral landscape, predominantly flat but interrupted here and there by steep escarpments that approach right to the riverside, like Cleeve Hill and Marlcliff. The villages are mellow settlements of well-worn stone buildings and the bridges mainly narrow structures dating well back into packhorse days. Near the bridges there is generally a pub or a village or both; but otherwise, the Upper Avon is rather a lonely river, and none the worse for that. It contrasts strongly with the lower reaches which are more beautiful, more fashionable and more densely used by boats.

The Lower Avon Navigation takes over from the Upper Avon at Evesham, above the lock. The difference is emphasized by the rather dashing A-framed house which straddles a side weir by the lock itself. The river's course through the town is lined by plenty of moorings and useful facilities for boats, while the waterside is tidy and well-groomed, giving the impression of a town more than happy to welcome those who arrive by boat. The river sweeps round the town itself in a long horseshoe, which takes it into a deep, well-defined valley with fruit orchards in the meadows on either side and rolling wooded hills just beyond. To the south is Bredon Hill, a huge green hump that dominates the river's course for many miles as the Avon wanders half way round this landmark. There are more locks—at Chadbury, Fladbury and Wyre Mill—which are noticeably different from those further upstream, being accompanied by a lockside house or large mill; and the weirstream is well used as a secure mooring for motor boats off the main navigable channel. Fladbury Mill is a large handsome pile, standing between the main weir and the lock, a few hundred yards downstream of an elegant railway bridge.

River Avon at Evesham, the start of the *Upper Avon Navigation*.

The old bridge at Pershore on the *Lower Avon Navigation*.

Further down is Wyre Piddle (with a riverside pub), and then Wyre Mill Lock; the mill has been usefully converted into a boat club. Pershore Lock is another sheltered lock but of curious construction. It has shallow but sloping sides, and boats using it must keep to the centre section. Further down still is Pershore, a beautiful old town cursed with a busy main road. The nearby bridge is yet another of the venerable Avon bridges which, with the mills, are distinctive and enjoyable features of the navigation.

Strensham Lock is at the head of the great pastoral reach that stretches down the west side of Bredon Hill to the village of Bredon. The latter is very pretty, almost extravagantly so, and crowded with boats. There is a large and famous old tithe barn here, but the idyllic rural scene is somewhat undermined by the presence of the M5 motorway. The slender, low-slung viaduct does its best, with minimum visual interference in this vulnerable landscape, but the noise is inevitable and continuous. Another long reach through rich, flat grazing land brings the river to within sight of the massive tower of Tewkesbury Abbey. This is the busiest part of the river, with sailing clubs, boatyards and marinas generating so much traffic that it begins to resemble the River Thames. At Tewkesbury, the Avon runs under the ancient King John's bridge, and weirs and a nearby lock drop the biggest part of the river into a side channel that joins the Severn nearby. The rest of the river continues under a low but decorative iron bridge as no more than a wide and navigable millstream, ending up at the clapboarded Abbey Mill. The centre of town is close by.

The River Avon has had a history more checkered than most inland navigations. It was made navigable before the middle of the seventeenth century by a private individual, William Sandys. By installing weirs and locks along the river he enabled boats to navigate, although it was by no means a waterway for deep-drafted vessels. The navigation below Evesham was later bought by one George Perrott and became the Lower Avon. The Upper Avon from Evesham to Stratford was owned by a group of business men, but the river as a whole was of little value as any form of long-distance trading route, for it was remote from any industrial area and served virtually no purpose other than that of a local agricultural waterway. Its course was winding, it had no useful connections (the narrow and heavily-locked *Stratford Canal* was of little use as a feeder waterway) and it was perhaps inevitable that the river should decline as a commercial navigation. Its northern end preceded the *Stratford Canal*'s decline, becoming virtually disused and unnavigable in the 1870s. The lower part of the river, which adjoined the important trade route of the far bigger and busier River Severn, survived much longer. Nevertheless, this too by the 1950s had become unnavigable above Pershore.

But by this time, the first stirrings of a movement to champion the inland waterways of this country had made themselves felt. By 1951, the whole of the *Lower Avon Navigation* had been purchased by C. D. Barwell who spearheaded the campaign of the (newly formed) Inland Waterways Association to restore the navigation. A Lower Avon Navigation Trust was formed, funds raised, and the river once again made navigable as far as Evesham (by 1963); and after it became apparent that the campaign led by David Hutchings to restore the *Stratford-upon-Avon Canal* from Lapworth to Stratford was steadily approaching a successful conclusion, the pressure for reopening the final link in the Lapworth–Tewkesbury waterway route became stronger than ever. So by 1968 David Hutchings had started work on the daunting prospect of completely rebuilding a river navigation which had been unused for a century. Inspired leadership and continual supplies of voluntary labour, including that of soldiers, prisoners and Borstal boys, tackled the endless problems—not all of which were merely physical. Legal obstructions raised by adjacent landowners and officious public servants greatly increased the amount of work that had to be carried out, including the construction of locks where none had existed before. It was indeed a very major piece of civil engineering, and yet it was achieved in five years at a cost of only £250,000.

The *Upper Avon Navigation* was reopened by the Queen Mother on the first of June 1974. With the boom of pleasure boating on inland waterways, the Avon faces a brighter future than at any moment in its 340 year lifetime so far. There is little doubt that the Avon will become as lively and as prosperous as the Thames—although it is possible that the final chapter in the Avon rebuilding story has yet to be told, for there is now a serious proposal to extend the upper navigation of the river past Stratford, to join the *Grand Union Canal* between Warwick and Leamington, following the river's course from Alveston Weir up through the villages of Hampton Lucy and Barford and past Warwick Castle. Such a plan, which only a few years ago would have been derided as absurdly unrealistic, is now a very real possibility. The two essential ingredients—enough money to start work, and the driving force of David Hutchings—are already available. The Avon has never before been navigable above Alveston; so if the Higher Avon scheme is carried out, it will represent a completely new scale of waterway restoration in this country, and such a breakthrough could bring a whole new group of abandoned rivers and canals into the possible-restoration bracket. The scheme would once again provide a wide waterway link between the Thames and Severn rivers—if the *Kennet and Avon* doesn't manage it first.

The river is an extremely popular waterway, far busier than the canals or its neighbour, the Severn. Most of the boats are very much of the small motor cruiser variety rather than canal boats. Up to now the river has been mostly heavily used towards its southern end and the outlet to the Severn, and although it is clear that boats are spreading upstream to take advantage of the recently restored waterway up to Stratford, the extensive facilities on the Lower Avon will inevitably continue for a while at least to hold most of the craft in those reaches. The south end of the river is extremely busy at weekends, with congestion at locks reminiscent of the Thames. Partly for this reason, there is a resident lock keeper at both Tewkesbury and Strensham locks. At the former, where the modern lock house is built on stilts because of the danger of flooding, the lock keeper sells temporary or permanent licences to boats wishing to navigate the river; the lock keeper at Evesham sells licences for the Upper Avon.

As far as the waterway itself is concerned, the river is easy and pleasant to navigate. The banks are easy to moor to, although unfortunately there are plenty of places where riparian owners forbid visitors. Some of the lock cuts are very narrow, requiring care on the approach, and when rain has raised the level of the river the current from the weirstream can be a slight problem. The locks are easy to use—many of them above Evesham are equipped with ex-Thames paddle gear, laid horizontally. Some of these locks are named after principal contributors to the river's restoration. Although the river is officially navigable from Tewkesbury to Stratford only, the three mile stretch from Stratford up to Alveston Weir is not obstructed by anything and can often be navigated by shallow-drafted cruisers. But boaters should accept the risk of grounding on the pebble bottom, and it is advisable to have either strong boathooks or to be prepared to jump overboard and heave.

The Avon bridges are worth a mention. They are not numerous, but they seem to make up for their scarcity by presenting to the navigator an obstacle he can hardly ignore. With the exception of the modern road bridges (including the slender M5 motorway bridge at Bredon) and the railway bridges (the Oxford–Worcester line crosses the river three times), great handsome structures that span the river effortlessly, the bridges over the Avon are almost all picturesque, venerable and doubtless much-cherished stone structures of hideously limited dimensions, considerably more difficult than many narrow canal bridges. This is even true of St John's Bridge in Tewkesbury, certainly the busiest reach on the river.

Stratford-upon-Avon Canal

The *Stratford-upon-Avon Canal* had a rather uneven start. Stratford was already linked, via the Upper Avon Navigation, to the Vale of Evesham and the Severn Valley; but by the end of the eighteenth century, the men of Stratford were keen to acquire a link to the narrow canal system then spreading fast throughout the Midlands. A link northwards would give them access to the cheap coal and manufactured goods of the Black Country in exchange for the produce of southern Warwickshire.

Like the *Worcester and Birmingham,* the *Stratford Canal* was designed with a single, very long summit level and a long, steep drop to the south. And just like the former, the Stratford company made a brisk start on building the summit level (in three years they reached from King's Norton almost to the top of the descent at Lapworth), then stopped short of the long and expensive programme of lock construction facing them. Hence the first eighteen locks and the short arm to join up with the new *Warwick and Birmingham Canal* were not ready until 1802; the rest of the canal down to Stratford was not even started until ten years later, and the line was not completed until 1816, having been finished off as quickly and cheaply as possible to get it at last open to trade. The little iron split bridges along this section are surely evidence of a good cost-paring idea exploited to the full, as indeed are the barrel-roofed lock cottages—charming as both these characteristics seem to us today.

The *Stratford Canal*'s lower section did not wear well. The Upper Avon was never as useful a connection as had been hoped and the canal, because of its very limited dimensions, numerous locks and unfavourable geographical position, was in no position to carry much goods, except traffic to and from the town of Stratford. The canal's proprietors sold out willingly to the Oxford, Worcester and Wolverhampton Railway, a forerunner of the Great Western Railway, and the canal declined steadily in usefulness as a transport route. Over the years, the northern section became infrequently used and the southern section eventually became quite unnavigable. By 1958, Warwickshire County Council had decided that the decaying canal should be formally closed, but the Inland Waterways Association fought the proposal and campaigned for its restoration. The National Trust became interested, a restoration project was initiated, and under the leadership of David Hutchings the canal was painstakingly restored. This involved the removal and disposal of countless thousands of tons of mud from the canal bed, while nearly all the lock chambers had to be rebuilt, re-gated and re-equipped with paddle gear. The labour force included soldiers and

Lapworth: the gap between the two iron 'leaves' of the bridge allowed the towing horse to cross from one side of the canal without the line to the boat having to be undone. There are several examples of this design on the *Stratford-upon-Avon Canal*.

prisoners and was almost entirely voluntary, and the work was carried out for a capital sum which was trifling compared to what it would have cost had it been done by a contractor. The canal was re-opened by the Queen Mother in 1964, and Hutchings promptly turned his attention to the River Avon below Stratford. This too has now been entirely restored—the final obstacle below the river lock at Stratford was removed and the whole river reopened in 1974.

The *Stratford-upon-Avon Canal* is usually treated as two distinct sections, the division neatly indicated by a junction half-way along. To the north is a twelve and a half mile section, lockless for all but the last one and a half miles. It is a waterway that starts in Birmingham's suburbs and runs through them for several miles, although the winding cutting that shields the canal maintains the illusion of a sheltered rural course. But away from Birmingham's furthest tentacles, the canal pursues an open, and more authentically rural if less spectacular course through Warwickshire.

The southern half, built a little later, passes through a much more snug and intimate countryside. This half is also heavily locked and contains a

196

string of features, many of which are peculiar to that stretch of water and to none other. It was the southern section of the *Stratford* whose restoration and reopening in 1964 gave heart to canal enthusiasts throughout the country. This canal now belongs to the National Trust and is jealously guarded by those who live along its banks. Because of this, and because everyone who contributed to the restoration project feels that he has a tiny stake in the waterway, the southern *Stratford Canal* has an atmosphere of being unusually well protected, cossetted even. Since 1964, the trip by boat down to Stratford-upon-Avon itself has been something of a pilgrimage by those determined to show that the canal's reopening had been worth it, and that no amount of locks (there are, on average, three per mile between Lapworth Junction and Stratford) would stop them going down to Stratford and back just to make the point. But now the canal has a more soundly based future in front of it. The reopening of the *Upper Avon Navigation* from Stratford down to Tewkesbury and the Severn has made the *Stratford Canal* part of a through route for the first time in a hundred years. Although this will inevitably increase maintenance costs (the canal is still run on a shoestring), the toll revenue should look considerably healthier.

THE COURSE OF THE CANAL

The canal's western terminus is a junction with the *Worcester and Birmingham* at King's Norton. Close to the junction is the unique King's Norton Stop Lock, a remarkable guillotine arrangement bracketing the first bridge on the canal; and nearby is the canal's only swing bridge, serving a small works.

Although the canal's summit runs through a built-up area, the waterway itself is completely screened from its influence by a useful curtain of vegetation. Brandwood Tunnel (sometimes known as King's Norton Tunnel) is not far away. Its western portal carries a carved stone head of William Shakespeare. The tunnel's cross-section is of comfortable dimensions, like the rest of the bridges on the northern section of the waterway—a reminder of the original high hopes for the canal as part of an important through route between the Dudley coalfields and the *Warwick and Birmingham*. The canal continues to wind secretly through several miles of anonymous built-up areas before passing under a railway and heading off south-east into a more realistic countryside, although suburbs or reminders of them are still to be found. At one point there is a partly navigable feeder bringing water down from Earlswood Lakes. Eventually, past a couple of steel bascule bridges, the long level pound that stretches back as far as Birmingham and Wolverhampton is ended with the start of the Lapworth flight of locks. From here, the canal's character changes drastically as it drops steeply down the first of fifty-five locks to the River Avon at Stratford. The locks lead the canal into an altogether prettier landscape of small lumpy hills, thatched timbered cottages and all the apparel of Shakespeare's country. The canal changes too, abandoning the guise of a reasonably broad waterway it displays on the northern pound, and becoming in every way a narrow canal of tiny proportions. If the locks are narrow, the bridges are minute, being only just wide enough to allow a seven foot boat to pass. Most of these are the iron split bridges peculiar to this canal, to be seen from Lapworth southwards. Lapworth itself marks the junction with a well-used arm to the *Grand Union Canal,* and is thus an important 'halfway-house' on the canal. The locks continue through the sheltered hamlets of Lowsonford and Preston Bagot, from where there are two level pounds of three miles each separated by a solitary lock (called 'The Odd Lock'). On these pounds are the canal's two principal aqueducts, at Wootton Wawen and Edstone. The latter carries the little-used railway branch to Stratford, which stays near the canal, though mostly out of sight, all the way down to Stratford. At Wilmcote, a further flight of eleven locks takes the canal down towards the outskirts of Stratford. In the town itself, the canal follows a narrow, hidden corridor, dropping steeply through a last handful of locks before suddenly opening out at the terminal basin in the heart of Stratford. A broad lock—the only one on the canal—connects the basin to the River Avon.

PLACES TO SEE THE CANAL

The northern section of the canal is attractive in a rather ordinary Midland kind of way, but it is inevitably the southern half that has so much character of its own. On the northern half, the stop lock at King's Norton is worthy of inspection, being a type which is otherwise unknown on the narrow canals. Its iron frame, pulleys, wheels and heavy chains are all rusted up now, and the wooden gates jammed in position above the water. The guillotine arrangement enabled the lock to be used as a two-way stop lock, with the gates facing neither one way nor the other but allowing any difference in water level between the two canals to be maintained.

LAPWORTH: There is a carved house at the junction, a barrel-roofed cottage next door, a lock at the entrance to the National Trust's canal with a roving split bridge over the lock head. At the tail of the lock is another very low bridge of permanently temporary appearance, a canal maintenance yard and the National Trust office (the southern canal is administered and maintained from here); there is also a boatyard. A few yards away are a pair of tiny feeder reservoirs, and a side lock down into a short branch that leads to the nearby *Grand Union Canal*. This shallow lock exists as a stop lock to protect the latter, which was built a little earlier. Apart from the cheerful array of converted narrowboats moored near the junction, the canal at Lapworth is dominated by locks more than by anything else. The top lock is one and a half miles to the west, but the flight starts in earnest little over half a mile from the junction. The locks are spaced very close together in places, the short pounds in between extending right along the lockside to maximize their capacity and enable them to supply at least one lockful before running dry. Further down the flight this technique was not necessary.

EDSTONE AQUEDUCT: Further down the canal, towards Wilmcote, is Edstone or Bearley Aqueduct. This very substantial yet elegant structure is an iron trough carried on brick arches, and crosses two railway lines and a minor road. Its height is relatively insignificant but the length of well over 500 feet makes it the second longest iron canal aqueduct in the country. The towpath runs beside the trough on the base level, and is edged by an iron balustrade. The Odd Lock is just to the north.

Other places worth seeing on this canal are Wilmcote, where there's a good flight of locks, and the old timbered farmhouse nearby where Shakespeare's mother used to live; and Stratford-upon-Avon, where the canal's terminal basin is beside the river. There are also some good pubs on or near the canal at Wootten Wawen, Preston Bagot and Lowsonford.

BOATING ON THE STRATFORD CANAL

It is obvious from the above that the canal is heavily locked from Lapworth to Stratford. Most of these locks are fairly shallow and therefore easily negotiated. The southern section is owned and run by the National Trust, and a licence to navigate it must be obtained from the office at Lapworth Junction. Boaters are asked to be especially careful with the lock gates and paddles, which the Trust finds extremely expensive to maintain. In addition, the water supply to the canal is very limited, and boaters are asked to close all paddles *and* gates.

Worcester and Birmingham Canal

Those who travel by train between Worcester and Birmingham will be familiar with the Lickey Incline—a long, straight and very steep bank

Emerging from King's Norton Tunnel on the *Worcester and Birmingham Canal*. As in most canal tunnels, the roof drips almost incessantly.

200

The 'guillotine' stop lock at King's Norton, by the junction of the *Stratford* and *Worcester and Birmingham* canals. It protected one company's water level at the expense of another.

between Barnt Green and Bromsgrove stations that takes the railway down 300 feet from the Lickey Hills and the Black Country plateau, to the first suggestion of the Vale of Severn. Going down the bank, the brakes are applied almost continuously; going up, the driver 'takes a run' at the bank, the diesel locomotive snorting and snarling away as the speed drops right down to a mere twenty mph or less before the train finally struggles to the top and picks up speed once more.

Just a mile to the east, the *Worcester and Birmingham Canal* tackles the same gradient in its own way—a series of locks that is known to all canallers as the longest flight of narrow locks in Britain: thirty locks in under two miles, with a further six in the next mile. It is a flight that appalls some boaters, but to certain others, who are convinced that working through canal locks is the most enjoyable form of manual labour yet devised by man, it is all they could wish for. The *Worcester and Birmingham* is not all locks, however.

201

The fifty-eight locks in the thirty miles between Worcester and Birmingham are all grouped in the southern half of the canal, leaving a fourteen mile lock-free pound from Tardebigge to Birmingham—although, as if in compensation, the pound features four tunnels of widely varying lengths.

And yet the canal is not just something out of the average in terms of lock and tunnel statistics. It is a waterway of great beauty during its course, not only through the hills and woods of Worcestershire but even in its passage to the heart of Birmingham, carried as it is in a leafy cutting right through most of the outskirts of the city, an approach quite unequalled anywhere in the Black Country. And to the navigator, one of the canal's advantages is that it was originally designed as a wide canal, so there is plenty of room at the tunnels and bridges. (By the time there was enough money available to build the Tardebigge locks (there are thirty) and complete the canal line, the canal's promoters had changed their minds in favour of a narrow canal.)

The *Worcester and Birmingham Canal* was designed as a short Birmingham–Severn route, linking the expanding manufacturing area of the Black Country with the great trade route of the River Severn. Designed as a barge waterway to enable the Severn Trows to reach Birmingham itself, the *Worcester and Birmingham* was obviously a grave threat to the *Staffordshire and Worcestershire Canal*, which would become a longer and much slower route than the new canal. The *Birmingham Canal* also felt itself jeopardized, and written into the *Worcester and Birmingham's* Act, passed by Parliament in 1791, was a clause forbidding the new canal to approach any nearer than seven feet of the *Birmingham Canal*. This was the origin of the Worcester Bar at Birmingham, an infuriating obstruction which, coupled with the difficulty of obtaining adequate water supplies to feed the *Worcester and Birmingham's* summit level, caused the company severe headaches.

The canal was an expensive one to build, especially the long level pound south of Birmingham. Major cuttings and embankments were required between Selly Oak and Birmingham to maintain a direct yet level route. This difficult stretch was completed first, in 1795, but it had already used up most of the company's resources, and the rest of the line, with the great drop down to the Severn, had still to be tackled. After experiments with lifts, and after considering the projects then in hand for joining two narrow canals (the *Stratford* and the *Dudley*) to their line, the company decided on a flight of conventional narrow locks from Tardebigge southwards. The canal was eventually completed to Worcester in 1815, having cost nearly £600,000. In the same year, the Worcester Bar in Birmingham was replaced by a stop lock.

The canal was never particularly profitable, although the company tried

An iron aqueduct at Bearley. Note the level of the towpath.

all sorts of enterprising measures to maintain the revenue level. They sponsored a short canal to join Hanbury Wharf to the *Droitwich Canal,* creating a short route up from the River Severn to tap the salt trade from that town. The *Worcester and Birmingham* also leased the *Droitwich Canal* itself; and the company later tried, with some success, to turn railway competition to their advantage. In fact, the canal at one stage was almost converted into a railway line. But it survived, and was later taken over by the next-door neighbour-but-one, namely the Sharpness New Docks Company which owned the *Gloucester and Sharpness Ship Canal.* But the new owners failed to make it pay, and tonnage declined over the years. However the canal remains intact to this day—although both of the Droitwich canals were abandoned in 1939. Today there is a good chance that one or both of the latter may be recovered; the frequent working parties of today's 'navvies' show the way ahead.

THE COURSE OF THE CANAL

The canal runs from Birmingham to Worcester in a generally south–south-westerly direction. It starts at Worcester Bar Basin, Birmingham—a reminder of the deliberate obstruction enforced on the Worcester and Birmingham Canal Company by the older-established Birmingham Canal Company. From the basin, the canal makes off to the south through a secretive course shared only by a railway line.

South of Selly Oak, the canal runs also through the Cadbury chocolate factory at Bournville, once a principal user of the canal for transport, and through an industrial belt at King's Norton where the *Stratford-upon-Avon Canal* comes in. Just to the south is the entrance to the long West Hill Tunnel which penetrates the ridge dividing the southern limits of the Black Country from the pleasing hilly countryside of Worcestershire—which encloses the canal all the way down to Worcester. It is a countryside of fruit trees, black-and-white timbered houses and rolling wooded hills, dense hedgerows and quiet sunken lanes—an uneven landscape which was of no help to the canal's builders. To maintain the canal on one level while avoiding the many hills and dips, they had to cut a very twisting line of waterway, in contrast to the rather more regular, straightforward route further north.

North of Alvechurch are two of the principal reservoirs built to supply water to the canal's summit level. One above the canal has a feeder that used to be navigable (to allow access to the coal boats that used to fuel the old pumping station). The other is beside the canal, at the foot of an embankment, and is a popular place for water birds. In the hilly country between Alvechurch and Tardebigge there are two tunnels of 608 yards and 568 yards in length. Between them is a boatyard, and at the south end of Tardebigge Tunnel is a wharf which, with its basin, canal maintenance

Tardebigge Top Lock on the *Worcester and Birmingham,* with Tardebigge church in the background. This is the deepest lock on the Inland Waterway System, with a fall of fourteen feet.

yard and moored boats, has become something of a canal centre. The Tardebigge flight of locks starts close by, a long steep descent winding down through a quiet and sheltered landscape. There is a small reservoir beside some of the upper locks, and the pumping station nearby was built to pump water up to the top of the flight. At the bottom, there is only a brief respite with, happily, a canalside pub, before another six locks are encountered, taking the canal down to Stoke Wharf. South of here is a small outpost of now-derelict industry—it used to be a salt extraction works beside the canal and once an important source of traffic for the canal itself.

Near Droitwich, the canal passes the *Droitwich Junction Canal*—one of the 'youngest' of the English canals (it was opened in 1853) but now derelict and overgrown. South of Dunhampstead Tunnel—the fifth and southernmost on the canal—the canal turns west, under the M5 motorway and down the six locks of the Offerton Flight. From here locks occur increasingly frequently as the canal approaches Worcester and the level of the River Severn. In Worcester itself, the canal follows a rather dingy course through the nether regions of the town, eventually ending up at the spacious Diglis Basin on the south side of Worcester. From here two barge locks drop down into the river.

PLACES TO SEE THE CANAL

The three principal points of general interest are noted below (the junction with the *Birmingham Canal* at Worcester Bar Basin is treated elsewhere). Other noteworthy places along the *Worcester and Birmingham* include the canal's passage through Cadbury's chocolate works at Bournville; the canal reservoirs at Bittell, north of Alvechurch; the old salt works at Stoke Prior and the nearby Stoke Wharf; the delightful hamlet of Oddingley, with its church by the canal; Offerton Locks, pretty in their open setting beside a farm despite the adjacent M5; and the big basins off the main line at Lowesmoor, in Worcester itself. There are plenty of canal pubs in useful spots too: like the Queen's Head at the bottom of the Tardebigge Flight, the Eagle and Sun at Hanbury Wharf, and the Fir Tree at Dunhampstead Wharf.

TARDEBIGGE: One of the focal points of the waterway, Tardebigge Wharf is at the top of the long, steep drop down to the Severn. It is here that boat crews tend to pause before taking a deep breath and tackling the locks; and conversely, it is at Tardebigge Wharf that exhausted north-bound crews often stop to lay up for the night. The basin and drydock near the tunnel entrance have long been for the purposes of maintenance staff, but, not surprisingly, quite a substantial canal settlement has developed. The top

Offerton Locks, *Worcester and Birmingham Canal*. The building on the right provided stabling for canal horses.

lock is not far from the wharf and is overlooked by a lock keeper's cottage. It has a fall of fourteen feet, which makes it the deepest narrow lock in the country. Originally a boat lift stood here, an experiment carried out at a very early stage by the canal company to see if heavy lockage down to the River Severn could be reduced by installing a series of deep lifts. This was ready by 1808, a fairly straightforward device incorporating a wooden tank suspended from overhead iron pulleys which worked fairly well; but the company was sufficiently confident of its summit level water supply to decide, in the end, to stick to the conventional pound lock system. Nothing now remains of the structure.

The Tardebigge Flight does not begin in earnest until the second lock; and then they come thick and fast, separated by pounds of only fifty yards or so. A reservoir lies beside some of the upper locks. Its water was pumped up to the summit level by the engine house nearby.

HANBURY WHARF: A centre now of pleasure boating activity, Hanbury Wharf was once an important transhipment point from the canal to the important salt-producing town of Droitwich, just one and a half miles west. Droitwich was connected to the River Severn by a broad canal, now disused. Traffic between this canal and the *Worcester and Birmingham* naturally had to be carted by land, until the construction of a link canal, the *Droitwich Junction Canal,* which was opened in 1854 between Hanbury Wharf and Droitwich. In spite of its late construction, this canal did not last very long and was officially abandoned in 1939. It is now overgrown although the flight of narrow locks can still be seen; the top lock is close to the junction.

DIGLIS BASIN, WORCESTER: A good example of a canal-river port, Diglis was once the scene of the bustling exchange of goods between narrowboats and Severn Trows working up the river from the ports downstream. All the trading craft have gone but the basin continues to thrive, its life now based on the pleasure boats of all shapes and sizes; and many of the old warehouses still survive and are still used. There are in fact two basins at Diglis, an outer and an inner one. Two barge-size locks lead down to the Severn. Since, in spite of side ponds, these drain more water out of the basin than lockfuls of the narrow canal could ever replace, electric (formerly steam) pumps lift water into the basin out of the river.

BOATING ON THE WORCESTER AND BIRMINGHAM

The *Worcester and Birmingham* offers cruising on an exceptionally beautiful waterway. That the canal has been well appreciated as such for a considerable time is proved by the number of pleasure boating centres that have been established along its banks for many years. Its northernmost section has anyway been long acknowledged as the most painless way of navigating to the centre of Birmingham from the south. And now that the Upper Avon Navigation has been rebuilt and reopened (forming part of a beautiful but heavily locked 110-mile circle of cruising waterways) the *Worcester and*

In the Tardebigge flight of locks. Fibre glass boats have to be handled at locks much more carefully than heavier, stronger, craft.

Birmingham and *Stratford-upon-Avon* canals find themselves busier than ever—which is no bad thing for all concerned. On the *Worcester and Birmingham,* the dense lockage from Tardebigge to Worcester often deters the faint-hearted from exploring the southern—and prettiest—reaches of the waterway. The locks are in fact less tiresome than might be suspected; they are well maintained and easy to operate, and their design allows for rapid emptying and filling. To work through the flight in pouring rain, however, is by no means everyone's idea of holiday fun. At Dunhampstead is based the Cedar, a traditional narrow boat which runs day trips from Dunhampstead Wharf. Boating firms are scattered randomly along the waterway between Hopwood (just south of West Hill Tunnel) and Diglis Basin, Worcester.

Monmouthshire and Brecon Canal

The 'Mon and Brec' is today's nickname for the beautiful canal in South Wales that runs along the eastern edge of the Brecon Beacons and down the Usk Valley from Brecon to near Pontypool. It is useful to have such a nickname because, theoretically, this continuous length of navigable waterway is called either the *Brecknock and Abergavenny Canal* (its original name), or the *Monmouthshire and Brecon* (which was formed by the *Brecknock and Abergavenny's* later amalgamation with the *Monmouthshire Canal,* which ran from Pontypool down to Newport Docks and, thus, to the Bristol Channel and is now abandoned). In fact, today's 'Mon and Brec' includes (just) a bit of both canals.

The old *Monmouthshire Canal* was a short industrial waterway serving local coal mines and iron works, with a network of early tramways. The *Brecknock and Abergavenny Canal* was built along the same lines, but its distance from the heavy industry near the coast gives it a dramatically different flavour. It is a thirty-three mile stretch of one of the most delightful canals in all Britain. Its character is very much defined by the mountainous setting and its 'contour canal' design. It follows the Usk Valley, holding a position above the valley floor and winding along the edge of the mountains. Overhung with trees, it hugs the hillside while its spectacular views contrast with the tiny scale of the stone bridges and the locks. It is irresistibly attractive, and those who have seen it invariably want to return.

The canal has only recently become a boating waterway again, and since the demise of the line to Newport, the *Monmouthshire and Brecon* has been entirely cut off from any other water route. After World War II, it lapsed quickly into an advanced state of decay and was restored to navigability only in the later 1960s. Fortunately for the restorers, there are only six locks— five at Llangynidr and one near Brynich Aqueduct, a few miles short of Brecon. But these had all to be fully repaired, and a fixed bridge at Talybont replaced by a bascule. Thus the canal is once again open and much used by the owners of small boats, particularly for day trips. It makes an ideal intro- duction to the very best that canals have to offer and is well worth a visit.

II

South-East Waterways

RIVER THAMES

RIVER LEE

RIVER STORT

WEY NAVIGATION

KENNET AND AVON CANAL

MEDWAY NAVIGATION

CHELMER AND BLACKWATER NAVIGATION

River Thames

The Thames is the best-known and perhaps the best-loved of all England's rivers. It is a river of beauty, grandeur and style, and is still, as it always has been, the undisputed focus of virtually every town and village it flows through—a rare attribute in this country, and due to the river's immense popularity as a pleasure waterway. As a navigation, the Thames is a superbly maintained river, with immaculate locks and lock gardens that are worlds removed from the strictly functional and often sorely neglected locks to be found throughout much of the canal system.

The river has several important connections with the inland navigation system. Its top end—since the demise of the *Thames and Severn Canal*—is a dead end, but the Thames forms the southern terminus of the *Oxford* and *Grand Union* canals, and thus makes an excellent circular cruising route for narrow-beam boats. The Thames is also the gateway to the *Kennet and Avon Canal* and, via the River Wey, to the *Basingstoke Canal*. Restoration schemes for both these waterways have been very much dependent upon this navigable access.

THE COURSE OF THE THAMES

The river is over 200 miles in length from its source in Gloucestershire to the end of the *Thames* estuary and the start of the North Sea, near Southend. The change in the river's character from a tiny Cotswold stream to London's

Near the upper limit of navigation on the Thames at Lechlade is the silted-up entrance to the long-abandoned *Thames and Severn Canal*. In the background is one of the 'roundhouses' peculiar to this canal.

highway and the bleak estuary that enters the North Sea is obviously a gradual one, but it is realistic to treat the river in four different sections: the unnavigable headwaters from Thames Head to Inglesham, near Lechlade; the navigable river from Inglesham to Oxford; the river from Oxford to Teddington; and the tidal river from Teddington to the North Sea.

The Thames rises close to the Foss Way south of Cirencester, very close to the old *Thames and Severn Canal*'s great tunnel through the Cotswolds at Sapperton. (A stone statue of a recumbent Father Thames was placed here in 1958 to mark the spot, but has now been moved to a safer position at Lechlade Lock.) The stream quickly gathers strength as it passes through sleepy villages in Wiltshire and Gloucestershire. The official start of the navigable Thames is at Inglesham, just a mile above Lechlade, a spot marked by the original eastern terminus of the *Thames and Severn Canal*.

From Inglesham and Lechlade to Oxford the river is fully navigable, but very different in character to the reaches below. It is a small, pretty and very local river, with none of the smartness that helps make it so busy further down. Indeed, it is in some ways more of an angler's river than a navigator's, and the river inns (three of which are called the Trout) are full of fishermen. The river itself is isolated, quiet and unchanging, wandering alone through an area of tiny villages and hamlets. The locks on the river—which are manned all the way—are relatively small. Many of them are accompanied by the old 'post weirs'—old-fashioned weirs composed of ranks of wooden paddles attached to wooden posts. It takes a certain skill and strength to

Temple Island, near Henley-on-Thames.

draw these paddles, and to replace them. Like everything else about the navigation works, the weirs and locks and gardens are immaculately maintained; the locks on the upper river are also mostly manually operated. Many of the bridges are small, single-arched stone structures; it is here that the riverside inns are found. Out of season, these are as unhurried and congenial as any canal pub: it is that sort of river.

Although the river's approach to Oxford is superb, with distant views of the city's many spires across the rich green meadows below Godstow Lock, much of the river's passage through Oxford itself is pretty dismal, being shut away at the back of the town among the gasworks and the railway lines. In the comparatively pleasant urban stretch above Osney Lock is the lowest bridge on the navigable Thames, allowing only seven feet and seven inches of headroom. Frequently cursed by the owners of larger boats, Osney Bridge ensures that the upper reaches of the Thames remain relatively quiet, uncrowded and unspoilt.

Below Oxford, the river's contribution to the landscape and its potential for pleasure is exploited to the full. The low banks, riverside inns, and nearby meadows and woods are all very much part of the river scene. But what underlies the attraction of the Thames valley is perhaps that the Thames has always been a fashionable river, and although in terms of navigation the river went through 'a bad patch' in the nineteenth century, the Thames valley has never been anything but a desirable area in which to live, for not only is it of great natural beauty but it is also accessible by water—an important consideration for the builders of Windsor Castle and Hampton Court. Thus, all the way along the river are large and graceful country houses built to use the river as a unique setting. Some of these, like Cliveden, are situated high above the water, although most of the others are built much nearer, with their lawns stretching unfenced to the water's edge.

But this river is not just a rich man's playground; it is also a focus for camping activities and a range of water sports. Apart from the bother of intensive use by motor launches, the Thames is a good sailing river and an excellent rowing one—Henley, Marlow and Pangbourne being three of the principal rowing centres. These towns, like virtually all the others along the Thames' course, present their most graceful and well-groomed fronts to the river. Hardly a whiff of industry intercedes (except for an unfortunately sited power station near Abingdon). But the elegance of the towns, the charm and variety of the bridges, the beauty of the Thames valley countryside and occasional highlights like Windsor Castle make a journey along the Thames a delightful voyage through the richest, softest parts of southern England.

All this comes to an end at the beginning of the tidal reaches of the Thames, at Teddington. Below Teddington the river is very much London's river, being large and wide, a broad commercial highway. It is certainly the capital's best geographical feature; it has also been the very base of London's prosperity since Roman times, and London still turns its best front to the river, with countless important buildings facing the Thames. The Thames bridges in London are worth a study by themselves—there are twenty between Hammersmith and Wapping alone.

Yet for all its advantages, the river in London is no longer used as much as it once was, from the point of view of both transport and pleasure. The amount of goods carried on the river through London is but a fraction of what it once was, and the number of river wharves still in use continues to fall. So too, up at Brentford, does the amount of Thames traffic feeding the *Grand Union Canal*. There is precious little commercial movement above Teddington these days.

Down at the other end of London is Dockland; once the biggest port in the world, now a shadow of its former self. The Pool of London rarely sees freighters now and Tower Bridge rarely opens except to admit ships on goodwill visits to the city. The Surrey Docks are silent and empty, and the *Surrey Canal* closed. The West India docks and the Royal Group of docks are still in business, but more and more shipping uses the docks down river at Tilbury. Thus by Tilbury the river has become a principal seaway although only half a mile wide. But then it broadens out into a great estuary and loses itself in the North Sea.

BOATING ON THE THAMES

The Thames is a very different boating proposition to most inland waterways. It will be obvious from the above that the non-tidal Thames is an ideal boating waterway in many ways. But this has of course made it very much *the* place to moor a pleasure boat, and the river tends to be extremely crowded. (Not long ago, there were more than twice as many pleasure boats on the Thames as on all the British Waterways Board's network of inland navigations.) This has led not only to congestion at locks and mooring sites, but also to a rash of 'private—keep off' notices and an inevitable lack of privacy; this applies less to the reaches above Oxford, where larger craft cannot penetrate.

The crowded condition of the River Thames in summer is a phenomenon which leaves other inland navigations looking like the forgotten waterways that they were but a decade ago. The idea of tying up in a lock for lunch, or even overnight—a forbidden luxury which some canallers are tempted to enjoy when the time and place is right—would be unthinkable on the Thames. Indeed, so great is the pressure of pleasure boats on the river that many locks have had to be rebuilt and greatly extended in recent years to remove the bottlenecks one by one—a slow and expensive process.

Taking a boat onto the river should not be done without some forethought, for there are strict regulations which must be met, and the Thames Conservancy officers reserve the right to inspect any boat to see for themselves. For example, there is an embargo on the discharge of lavatories or indeed any dirty water into the Thames—a measure which is largely responsible for the improving quality of the river water, at least above Teddington. Other rules deal with the design and layout of fuel and gas fittings, and most other aspects of boat design which could affect safety on board, as well as lesser points such as the display of the boat's name. All these are vigorously enforced by the Thames Conservancy and since many canal boats are normally subject to few such controls on their home waters, owners are advised to discover the Conservancy's requirements well before they reach the river. All boats should in any case be registered with the Thames Conservancy *before* joining the Thames, an element of orderly conformity which is very much the way of things on this river. It can be aggravating, but there is obviously good sense behind the rules.

Navigating the river in winter is of course a completely different experience. Hardly a boat stirs the water, save the occasional maintenance craft. The lock keepers are still on duty; even when there is nothing moving on the river they have to be ready to man the weirs at the first hint of floodwater, and are usually more than pleased to see a boat to break the monotony. In flood conditions, the Thames is as tricky to navigate as any other river; but on a crisp winter's morning, with the river as flat and silent as the proverbial millpond, a trip along the Thames is an exhilarating experience.

On the tidal Thames boating is very different, and the river is used much

Locks on the River Thames are immaculately maintained and efficiently operated. Most are hydraulically powered now, unlike the lock at the right in this photograph. Ever-growing numbers of pleasure boats in the reaches between Oxford and Teddington have compelled the drastic enlargement of several locks to remove bottlenecks.

more for sailing and rowing activities than for cruising. There is a certain amount of commercial traffic to the London wharves—but a much greater hazard to the pleasure boats is the large amount of driftwood that is carried up and down on every tide. Nor are things made easier by the conspicuous absence of landing wharves or jetties in London. Most of those that do exist are either reserved for the summer trip boats that ply along the river or are semi-private jetties administered by the Port of London Authority and available only for their own boats, and for those of the police and a few privileged others.

River Lee

North of London, the counties of Hertfordshire and Essex are separated by the valley of the River Lee—a valley which is typical of neither county yet which seems to incorporate some of the worst of both. Gravel pits, reservoirs, greenhouses, power stations, railways stations and caravan sites abound in the confused linear jungle of London's outer environs, a confusion to which the Lee Valley Regional Park Authority has, over recent years, been struggling to bring some form of ordered environmental improvement.

The focus of the valley is, of course, the River Lee itself, an ancient navigation which reaches up from London's dockland to the elegant milling towns of Ware and Hertford. Part river, part canal, the navigation cuts straight through the gravel pits and reservoirs in a business-like and most unromantic fashion. But it is not a consistent river, for its character changes forcefully from the cosy little river of Hertford and Ware to the bustling, industrial waterway of East London.

The Lee—or Lea—has been navigable for many hundreds of years. The first actual legislation concerning the river's improvement was passed in 1425. It has always been an important waterway in terms of supplying London with produce ('Ware—the granary of London' was the familiar slogan) in exchange for manufactured goods. (This role has obviously been modified, but grain barges may still be occasionally seen loading at Ware.) A later use for the River Lee was to supply water for the New River, the drinking water supply canal engineered by Sir Hugh Myddleton and opened from Amwell Spring (near Ware) to Islington, London in 1613. (The Lee was shortly afterwards tapped as a supplementary source.) The New River is another early piece of 'water engineering' of great significance and interest, but a detailed description of the New River has no place here.

The River Lee rises north of Luton, drops south-east through that town and feeds the ornamental lakes at both Luton Hoo and Brocket Park, taking in numerous water mills as well as the towns of Harpenden, Wheathampstead and Welwyn Garden City before joining the Mimram at Hertford. The head of the navigation is in the centre of the town. Between Hertford and Ware, the Lee flows alone through rich green water meadows. The New River is abstracted from this length, and indeed the Metropolitan Water Board—the successors of the New River Company—administer the lock at Ware. The river here adopts a more southerly course as it heads towards London. At St Margaret's, there are some old maltings beside the navigation, and not far from here the River Lee is joined by the River Stort as it heads south again—into a very different landscape, which features the uneasy co-existence of caravan sites and power stations, anglers and pleasure boats. A railway borders the west side of the valley, and huge power cables march along just to the east; the old gravel pits which have transformed the valley floor on either side are gradually being turned into something useful. In the distance are the greenhouses from which come the Lee Valley cucumbers and tomatoes.

South of Enfield Lock the navigation is lined by miles of reservoirs on one side and more and more industrial buildings opposite. It soon loses all trace of its more open upper reaches and becomes an increasingly enclosed industrial waterway.

The Lee's value as a short but direct link between London's dockland and what has become a prime manufacturing area is reflected in the very heavy commercial traffic that uses the navigation's lower reaches; and this determines very much the character of the river and its surroundings. On any working day, the navigation is alive with Thames lighters and tugs clanging about and lurching into lock gates, moored craft or anything else that gets in the way. It is a busy, noisome river from the Thames right up to Enfield. Apart from general merchandise and coal, the principal traffic would appear to be timber, and barges heaped with stacks of it serve wharves all the way from Old Ford to Edmonton. (The latter is an important manufacturing centre for furniture.) This explains why the river is sometimes completely covered with a layer of flotsam—a hazard more generally associated with the tidal Thames. Other characteristics of the Lee in London are the paired, electrically operated locks which help to keep the traffic moving. It is in every way a business-like waterway, and remains so all the way through Ponders End, Tottenham, Hackney and Bow.

The River Lee joins the Thames at Blackwall via the well-named Bow Creek, a twisting tidal waterway. Bow Locks are at the head of this creek, but most traffic uses the Limehouse Cut. This used to bypass the Regent's Canal Dock and lock down separately into the Thames, but this exit has now been closed and the Limehouse Cut diverted into the old Regent's Canal Dock. The lock into the Thames is just under two and a half miles downstream of Tower Bridge.

BOATING ON THE RIVER LEE

Although the upper reaches of the river are quite pleasant boating waters, there is no denying that the lower parts are pretty grim and mainly used by commercial traffic. There is little but riverside industry and occasional power stations along these stretches, while further north the long high banks of the reservoirs constrict the view.

The locks on this stretch are hydraulically operated by lock keepers. The gates and paddles are designed to work very fast, so it is unfortunate if the lock keeper takes a dislike to you and makes free with his machinery! North of Enfield Lock there is little commercial traffic, and boaters should be prepared to work themselves through alone. Some of the gates take quite a battering from the occasional barge and tend to leak badly.

Other things to watch out for are the weir pools at places like Dobbs Weir, especially in time of flood; and the heavy baulks of timber which are often floating on the lower reaches of the navigation. The worst section for this is just north of the Limehouse Cut—which is also extremely winding, narrow and busy. A good lookout is essential—and bear in mind the 'iceberg' principle. What appears to be an insignificant little plank is often something massive and very solid which can damage a propeller or even the hull of a boat.

River Stort

The Stort is a remarkable contrast to the Lee, of which it is the principal tributary. The river becomes navigable at Bishop's Stortford, and flows for just under fourteen miles before meeting the River Lee at Feilde's (sic) Weir, near Hoddesdon. Almost all of the river's course forms the county boundary between Essex and Hertfordshire. It is a beautiful little river of watermills and water meadows untarnished by any of the industrial surroundings and massive reservoirs that hem in so much of the River Lee. A remote cul-de-sac of the waterway network, the Stort is well worth exploring.

The navigation dates from the eighteenth century, when the river was improved in order to carry malt and grain southwards to London. Never a profitable waterway, it would doubtless have been closed long ago had it been only a canal. But although the river no longer carries grain traffic (not perhaps surprising on such a small river) it is ideal for pleasure cruising.

The head of the navigation at Bishop's Stortford is inauspicious, at the back of a municipal car park. However once out of the town the real character of the Stort is soon established—a pleasant low-lying river valley punctuated by occasional neat weather-boarded villages, like Sawbridge-worth and Roydon. It passes the new town of Harlow, and is none the worse for it. But the main features on the river are the mills that stand beside many of the locks. From the north downwards, the mills are at Twyford, Little Hallingbury (the long mill-stream forms a navigable branch and is used for moorings), Sawbridgeworth, Harlow, Parndon, Hunsdon and Roydon, an impressive string of mills by the standards of any navigable river. Unfortunately, none grinds corn anymore (Sawbridgeworth was the last to close) but at least none is derelict either. The one at Harlow is now a restaurant; Parndon Mill is a pottery workshop and gallery; and Little Hallingbury Mill is a boating centre. They are a mixed bunch, but most are the tall, white weather-boarded structures often found in Essex and Suffolk.

The River Stort is very painless from the navigation point of view. It is a minor river, and there are no dangerous weirs to watch out for, most of

The River Stort. Many of the locks on this navigation have a water mill beside them.

the flood water going down mill-races. However, in flood conditions progress is often halted because of reduced headroom under the bridges, which are very low to start with. The locks are all manually operated; lock keepers are found at some of them but it is sensible to adopt a do-it-yourself approach.

Wey Navigation

The River Wey was turned into a navigable waterway by an Act of Parliament of 1651, and was opened from the Thames to Guildford in 1653. The navigation was well ahead of its time, being equipped with turf-sided pound locks instead of flash weirs, and with a considerable length of pure canal cut keeping the navigation away from the river channel. A hundred and ten years later, it was extended up to Godalming, and this led to the successful promotion of the *Wey and Arun Junction Canal,* which provided a through waterway route to the south coast. All of this helped to boost the traffic carried on the Wey itself.

The main function of the navigation was to carry timber and agricultural produce down to London, receiving in exchange manufactured goods— and grain to be ground at the several mills along the river. The grain traffic has lasted—in horse-drawn barges—right up to modern times. The last run, to Coxes Mill at Weybridge, took place in 1970. Today, the river navigation is for pleasure boating only—for which it is ideally suited. The navigation is owned and cared for by the National Trust to whom the Wey from Guildford downwards—the original navigation—was transferred in 1964 by its previous and most benevolent private owners. Later, the Trust was also given the Godalming Navigation—the Wey from Guildford up to Godalming.

The Wey Navigation is neither a long nor a particularly dramatic one in terms of its structures and surroundings. It is, rather, a quiet and dreamy waterway, idling through the soft countryside—and outer suburbs—of Surrey. It is neither a rural nor an urban waterway since it manages a sort of middle course, just managing to hold towns and villages at arm's length. But here and there in the meadows and commons that accompany the navigation through the sheltered valley are one or two groups of buildings that bring the scenery to life—the spire of a church, a group of mill buildings straddling a weirstream, a waterside inn, a lock cottage and well-kept garden. There are fifteen locks, including two flood locks, and they occur evenly along the length of the navigation. All these features give the Wey Navigation a comfortable, homely character.

At Guildford, the navigation plunges into the centre of the town, and is an interesting passage by boat. Some inspired planning has sited the modern Yvonne Arnaud theatre on the lock island here. South of Guildford, the steep tree-clothed hills of the North Downs close in, guiding the river up to the head of navigation in Godalming.

BOATING ON THE WEY NAVIGATION

The navigation is well attuned to pleasure boating and is well worth an exploratory cruise. The entrance from the Thames is just below Shepperton Lock, where there is a maze of backwaters: the navigation is in the middle of these, and is fortunately signposted. A stop gate is installed below the first lock (Thames Lock) for use if the level of the Thames is too low for boats to clear the bottom sill of Thames Lock. Visiting boats can buy temporary lock passes here, and get fixed up with a windlass (1 inch spigot) if they are not already so equipped. Most of the locks on the navigation are un-attended. The Wey Navigation is best avoided after long periods of rain, as the river is prone to flooding and it is unwise to try and use a pleasure craft under such conditions.

Two canals that lead off the River Wey are worth mentioning here, although neither is officially open to navigation: the long-closed *Wey and Arun,* and the reviving *Basingstoke Canal.*

The *Wey and Arun* was authorized in 1813 as a broad, eighteen and a half mile long canal to link the River Wey at Shalford (between Guildford and Godalming) with the head of the Arun Navigation at Newbridge, near Billingshurst in Sussex. From here, the Arun Navigation continued the line

Even derelict canals are useful in places. These are houseboats on the *Basingstoke Canal,* in the heart of commuter-land.

for twenty-six and a half miles down to the sea at Littlehampton. A lateral branch from the Arun led to Chichester Harbour, with a subsidiary cut into Chichester itself. (Another navigation offshoot of the Arun was the River Rother, made navigable up to Midhurst in 1794).

This route from the Wey to the Arun meant, effectively, a broad water route right the way from London to Portsmouth—a route which gave rise to some ambitious schemes for a ship canal, although none of these came to fruition, and in the meantime the existing through route was severed when the *Wey and Arun,* a victim of the railway age, was abandoned in 1871 after a working life of only fifty-five years. The canal has been decaying ever since and has in places virtually disappeared without trace. Undaunted by the canal's sorry state, a group of enthusiasts have formed a Wey and Arun Canal Society with the declared aim of restoring to navigation the canal and its link with the south coast. Regular working parties now take place on the old canal, which runs through the attractive wooded uplands of Surrey and Sussex.

The *Basingstoke* used to run for thirty-seven miles from the Wey Navigation at Byfleet to Basingstoke itself. Closely followed by a busy railway (the London–Basingstoke line), the canal passes through Woking, and then through a landscape of pines and heathland that is more concerned with the military than anything else. The waterway goes right past the firing ranges of Bisley, as well as the barracks, parade grounds and general training area of Pirbright and Aldershot and, later, the airfield at Farnborough. The west end of the canal takes in delightful Hampshire countryside and places

221

The *Basingstoke Canal* near Woking, sadly in need of help.

like the picturesque town of Odiham. In between is a long pound twisting along the contours of the rolling countryside. The level is punctuated not only by the sturdy brick bridges and several wharves—at one of which (Ash Wharf) there was a barge-building yard on the canal bank—but also by unexpected and irregular 'flashes', where the canal temporarily widens right out to form a small lake.

The canal was opened in 1794, and had a checkered history during which it was continually being sold to new owners. It was last auctioned in 1949 for £8000, since when the ailing canal has become even more run down. The twenty-nine locks, of which all but one are grouped at the eastern end of the canal, have fallen into disrepair. Collapse of the 1230-yard Greywell Tunnel towards the western end had already severed the canal from there to Basingstoke, and much of this isolated length has been abandoned or filled in and is now beyond redemption. However, the rest of the canal is by no means abandoned, and the actual 'track' is more or less intact. Most of the locks need at least partial restoration, and the bed of the canal, which has suffered from indiscriminate rubbish dumping in many urban stretches, needs clearing and dredging out over many lengths. But the bed has not been filled in and the bridges have not been 'dropped', and nothing irrevocable has been done to the waterway which would prevent its restoration to full navigability from the Wey Navigation up to Greywell Tunnel.

The canal now has a useful and energetic body of supporters. The Surrey and Hampshire Canal Society has been actively campaigning for the canal's complete restoration up to Greywell, pointing to the great amenity potential of the canal in an area already crowded by urban sprawl. The two county authorities concerned, Surrey and Hampshire, have shown an enthusiastic and enterprising attitude towards the proposal to restore the waterway, and have applied compulsory purchase orders to remove the canal from its private owners, who were not interested in making the canal into a navigable waterway. It is now probable that the canal will be restored to navigation throughout its length east from Greywell. Already the canal society has restored the solitary Ash Lock, thus linking again the two western pounds that make up most of the canal's length. The next few years will doubtless see intense activity on the twenty-eight locks at the eastern end to reopen the link with the Wey.

The canal is certainly a worthy target for restoration efforts. Once restored, its secluded and tree-lined waters will not only extend the limits of the national, non-tidal network of waterways but will provide an ideal mooring for local boatowners who would normally be restricted to the uncertain waters of the River Wey or the crowded—and expensive—River Thames. The wide locks and the substantial headroom of the bridges will be particularly convenient in this respect. And non-boatowners will doubtless be happier to see the local dirty old ditch cleaned out and turned into a lively and worthwhile asset.

Kennet and Avon Canal

The *Kennet and Avon* is famous among canal enthusiasts because it is the subject of probably the biggest canal restoration project ever attempted in this country. Eighty-six and a half miles long with 106 locks, the *Kennet and Avon* is being slowly and painfully restored, lock by lock, from a state of decay over virtually the whole of its length—the result of many years of commercial decline and official negligence. The story of the fight to rescue the canal and drag it back to a fully navigable condition is a saga stretching right back to the last war, and it is by no means finished yet.

The *Kennet and Avon* has its roots in two separate river navigations—the Bristol Avon, and the River Kennet. The former is the river that flows through Bath, down to Bristol and along the Avon Gorge to Avonmouth

The *Kennet and Avon Canal* in Newbury.

and the Bristol Channel. It is a river whose lower, tidal reaches have always been navigable up to the old seaport of Bristol. Early in the eighteenth century the river was made navigable for a further eleven miles up to Bath, a town which was by then becoming a fashionable spa. At about the same time, way over in Berkshire, local interests promoted the making navigable of the River Kennet from the Thames at Reading up to Newbury. This became the Kennet Navigation which, in its eighteen and a half miles, incorporated several lengths of new canal cut to bypass sections of the river channel.

Although ideas of linking the Bristol Avon to either the Thames or the Kennet had been floated since the seventeenth century, it was not until the late 1780s that the 'Western Canal' was actively promoted. In 1794 the Kennet and Avon Canal Act was passed, authorizing the cutting of a line from Newbury to Bath that had already been surveyed by the engineer John Rennie. Completion of the fifty-seven mile canal in 1810 gave a direct water route from London to Bristol.

The Kennet and Avon Canal Company gained control of both the Avon and Kennet river navigations at an early stage, and since then 'The Kennet and Avon Canal' has come to mean the entire line from Reading to Newbury, Bath and the start of tidal waters at Hanham, just upstream of Bristol.

The canal runs through Berkshire, Wiltshire and Somerset, from the deserted water meadows and peaceful parkland of the Kennet valley up to the chalky hills of the Marlborough Downs, over into the Vale of Pewsey and on through the narrow confines of the Avon valley down to Bath and Hanham. There is a whole series of unspoilt villages along the route, and

224

A turf-sided lock on the *Kennet and Avon*, cheap to build but wasteful of water and inconvenient to negotiate.

west of Reading the only towns encountered—Newbury, Hungerford, Devizes, Bradford-on-Avon and Bath—are all of great charm. There is virtually no industry at all along the waterway. Framed by such beautiful surroundings as these, the canal can hardly help being an attractive one.

The length between Reading and Newbury comprises the old Kennet Navigation—the turf-sided locks give away its age. The river flows through gravelly water meadows and is crossed by numerous swing bridges. The navigation remains within the Kennet valley up to Hungerford, from where the canal turns away to climb steeply up to the summit level at Savernake, at which point stands the Crofton Pumping Station, one of the canal's unusual features. Inside this building, the two steam engines—one is an 1812 Boulton and Watt beam engine—have been restored to full working order and are steamed on open days for visitors.

From Crofton, the short summit leads through the broad Bruce Tunnel to the top of Wootton Rivers Locks—the start of the long drop down to Bath. But the descent is much less regular than in the Kennet Valley. The fifteen mile pound (the 'Long Pound') through the Vale of Pewsey provides a substantial boating waterway by itself. It gets weedy through lack of use, but this does not prevent the paddle boat *Charlotte Dundas* from churning up and down to show people the canal. At Devizes, beyond a series of elegant canal bridges, is the first of the twenty-nine Devizes locks, which include the sixteen Caen Hill locks that come in rapid succession. A large rectangular block of side pounds stretches out beside them. More locks at Seend and Semington bring the canal to Bradford-on-Avon and the entrance to the beautiful Limpley Stoke Valley. Two outstanding aqueducts—Dundas and Limpley Stoke—decorate this valley, and there is also the Claverton Pumping Station, an unusual waterwheel arrangement

225

Part of the historic machinery inside Crofton Pumphouse.

designed to feed water into the canal from the Avon, forty feet below. Like Crofton, this too has now been restored. From here the canal follows a glorious route into Bath before locking steeply down into the Avon. The Widcombe Locks have now been restored, two of them being joined

The *Kennet and Avon Canal* near Devizes. The 'track' of the canal is in good condition for most of the length of it. But restoration is slow because of the high number of locks that need to be completely re-gated.

together into a single very deep one. From here to Hanham and the start of tidal waters, the *Kennet and Avon* is entirely a river navigation.

Although the canal is not at present navigable throughout, the *Kennet and Avon* is in good hands. The Kennet and Avon Canal Trust Ltd is dedicated to the restoration of the canal and has already achieved much. Due to the Trust's efforts, the gaps between the navigable sections of the canal are narrowed every year. The biggest part of the work is probably the renewal of locks—a costly business when the new gates have to be made from scratch. There is also clearing of the canal bed to be done, as well as dredging and weed-cutting. On a different level, negotiations are necessary to try and solve the problem posed by old swing bridges, now fixed and carrying substantial traffic. Highway authorities are naturally reluctant to see a reopened navigation force a return to the delays and dangers caused to traffic on busy minor roads by the functioning again of the canal's swing bridges. But the most daunting physical obstacles are the dry length near Bradford-on-Avon and the twenty-nine derelict locks at Devizes. The latter will probably be left until last.

For a largely unnavigable canal, the *Kennet and Avon* is remarkably full of life. Its banks are constantly alive with people walking and fishing. Large and small boats use the weedy waters even in the isolated pounds, and the Trust's maintenance boats are usually at work somewhere, just as there is always a working party rebuilding a lock or clearing the towpath of under‑growth. At Easter there is the famous Devizes to Westminster canoe race. It seems certain that in the not‑too‑distant future motor boats will be able to reach from either end to Devizes, and perhaps right the way through. It is certainly a canal that deserves more than any other to be restored.

Medway Navigation

Round the corner from the mouth of the Thames at the Isle of Grain is the estuary of the Medway, its entrance guarded by the old naval town of Sheerness. The estuary is an ill‑defined river incorporating a profusion of low‑lying marshy islands, and mud flats and creeks on either side as well as, on the north bank, a crop of modern jetties both long and short serving the cluster of oil tanks that represent the great refinery on the Isle of Grain. It is this estuary that has made Rochester and Chatham the seafaring towns that they still are. But the Medway is navigable for a much greater distance inland than this. The tideway extends almost to Maidstone; and above the town the river continues as a fully navigable inland waterway through nine barge‑sized locks, right up to Tonbridge—sixteen miles above Maidstone.

The river has always been navigable up to Maidstone. But its extension to Tonbridge in the mid‑eighteenth century facilitated the transport of hops and other agricultural produce down the river—including timber for the dockyards of Chatham and supplies for London. For the latter traffic, there was once a very useful short cut—the seven‑mile long *Thames and Medway Canal,* which ran from the Medway at Strood and Rochester to the Thames at Gravesend, thus providing a link between the tidal but reasonably sheltered waters of both rivers. The *Thames and Medway,* which included a tremendous tunnel over two miles long through the chalk escarpment that divides the two estuaries, was converted into a railway line in 1845, only twenty‑one years after its opening as a canal, and since then the Medway Navigation has been an isolated waterway connecting only with the sea. This isolation explains why the upper Medway is little known outside its immediate neighbourhood, for few inland cruising boats are suitable for the sea passage from the Thames round the Isle of Grain and up to Chatham.

So the Medway has a life of its own, an uncrowded and unspoilt water‑way. It cuts through the North Downs, and its surroundings have a defi‑nitely Kentish flavour. Characteristics of the navigation include several very old arched stone bridges, and an unusual number of tributaries, which help to swell an essentially minor river into a broad tideway in the distance of only a few miles. The river also has a railway line as a close companion for much of the way; but this does not detract much from the scenery, and the numerous stations are very useful for people who wish to take 'one-way' walks along the river banks.

South from Rochester, the tideway extends to Allington Lock, just below Maidstone. The upper tidal reaches are framed by the hills of the North Downs: a large and messy cement works at Snodland and paper mills at New Hythe give the valley a distinctly industrial flavour. This continues above Allington Lock into Maidstone, where there are some long-established wharves and riverside works, but upstream of Maidstone the river becomes a very different waterway—narrow, rural, intimate, and very picturesque. Along the steep hillsides that enclose the river initially are the first of many hopfields which are overlooked in several places by some fine church spires. There are modest villages, a handful of inns and the low stone bridges. At Yalding the Medway is fed by water from the rivers Beult and Teise, and from here to Tonbridge it is a much smaller river, narrow and shallow as it twists a solitary course past orchards and hopfields up to Tonbridge. There are several locks along these upper reaches. The head of the navigation is in the centre of Tonbridge. In 1829, a private scheme to extend the navigation by six miles up to Penshurst was initiated; but the project was abandoned after work had begun so, unfortunately for today's navigators, the glories of Penshurst Place remain virtually inaccessible by boat. However, although Tonbridge is the theoretical limit of navigation, there is no physical obstacle to prevent craft venturing upstream for a mile or more to the second railway bridge. But beyond here the river is strictly for angling, and to continue upstream with a light craft can prove unpopular.

BOATING ON THE MEDWAY

The non-tidal navigable reaches of the river run to only eighteen miles, but they make excellent cruising waters. The bulk of the river, from Maidstone to Tonbridge, is vested in the Southern Water Authority, and the Authority charges a modest fee for the use of the locks; a lock pass can be obtained at the several boat businesses along the navigation or at the two manned locks (East Farleigh and Yalding). The rest of the locks are operated by boat crews: the lock paddles take a one inch square windlass.

The Medway, a delightful river in summer, is no place to be in wet winter weather. The water level can rise dramatically, and since the banks between Tonbridge and East Peckham are fairly low to start with, the river often floods, particularly in the low-lying area between Tonbridge and East Peckham. By this time the river will be flowing very fast and boating is not only difficult but dangerous.

Chelmer and Blackwater Navigation

The Chelmer and Blackwater is a short and little-known river navigation tucked away on its own in East Anglia. It serves to give Chelmsford an outlet to the sea. The navigation was opened in 1797, and the easy access it provided to the sea—and thus to distant markets for the region's agricultural produce—was very much instrumental in turning Chelmsford into a large and thriving town. The navigation was designed for wide but shallow-drafted barges, and it is worthy of note that the navigation was regularly used by such barges until 1972, when the last timber traffic along the full length of the waterway finally ceased. The navigation no longer carries any commercial traffic, but it is used for industrial water supply. There is at the moment little pleasure boating on the waterway, except for canoeing.

This navigation is an odd mixture of river and canal. Most of the route is formed by the River Chelmer, a small river down to Chelmsford that is then swollen by the two other minor rivers, the Can and the Wid. Apart from the terminal basin in Chelmsford, the navigation down almost to Maldon is a river navigation, with occasional—and very short—lock cuts and weirs every mile or so. It is a sleepy little waterway, wandering alone through the quiet, tree-fringed water meadows of rural Essex. Above Maldon, the Chelmer and Blackwater rivers converge, and the navigation follows a short link canal to abandon the Chelmer and join the Blackwater for just a mile round the back of Maldon. Then a length of canal leads down to the last lock at Heybridge Basin. In the meantime, the two rivers have joined up under the name of the Blackwater, which then widens out into a broad, shallow estuary, its waters merging imperceptibly with the miles of marshes.

Heybridge Basin is perhaps the most interesting point along the navigation. Although it is no longer the busy transhipment centre it would once have been with cargoes being transferred to and from tall sea-going sailing barges and the smaller canal boats, the basin is still very much in use, its calm, safe waters harbouring a variety of sea-going craft. The old sailing barges, the lockside weather-boarded cottages, the clatter of rigging and the whiff of the sea give this canal terminus an unusual atmosphere.

12

South-West Waterways

EXETER SHIP CANAL
BRIDGWATER AND TAUNTON CANAL
GRAND WESTERN CANAL

The *Exeter Ship Canal* is one of the very oldest canals in Britain; ironically its survival until the present day contrasts favourably with the many other canal and river navigations scattered throughout the West country—most of which have long been abandoned and are fading away into the fields from which they were cut. The *Grand Western Canal,* of which only the Tiverton Branch has survived intact, and the *Bridgwater and Taunton Canal* might once have formed part of a sea-to-sea canal that was projected to bypass Land's End; but in the event neither became any more than an insignificant and largely irrelevant length of navigation. But both still hold water, and the *Grand Western* has now been restored as a motor-boat-free country canal.

Exeter Ship Canal

This canal is one of the very few in the West Country that survive in a navigable state. It is also one of the very earliest in England, dating right back to the time of Queen Elizabeth the First. Indeed, the original locks built in 1567 were probably the very first pound locks in England. Up to then, and indeed elsewhere for many years afterwards, changes in level on a river navigation were overcome by 'flash weirs'. Ordinary canals as we know them, independent of river courses, did not at that time exist. (The *Exeter Canal* was built as a bypass to a weir on the Exe river. It was, and is, fed with water entirely from the Exe.)

The navigation today is considerably larger in its cross-section, being now a waterway big enough for 400 ton ships and with a depth of between ten and fourteen feet. This contrasts with the sixteen ton barges that were as much as the canal could handle with its three feet depth in the sixteenth century. It is also considerably longer than it used to be, even if that makes

The *Exeter Ship Canal.*

it only five and a half miles from end to end. The canal was still in com‑
mercial use until the early 1970s. Until that time, sea‑going tankers still
occasionally penetrated to a wharf half‑way along the canal, and there was
a small timber trade up almost to the head of the navigation. Now, con‑
struction of the M5 motorway bridge has ensured that the *Exeter Ship Canal*
can never be a proper ship canal again.

The head of the canal is a large basin half a mile below Exe Bridge in
Exeter itself. This basin has found a modern role as a home for the famous
Exeter Maritime Museum. At the other end of the canal is Turf Lock,
more than half way between Exeter and Exmouth. By this stage, the River
Exe has become a mile‑wide estuary, although most of it is extremely shallow
at low water. Between these two termini the canal runs along the bottom of
the valley between the main railway line and the river—a somewhat in‑
glorious stretch taking in the gasholder end of the city. But further down
towards Topsham and the estuary, the canal's surroundings improve con‑
siderably. There are not many bridges on the way: a couple of small swing
bridges, a large bascule lift bridge at the Exeter Bypass, and now the big
M5 bridge between here and Topsham. This motorway bridge is the only
fixed bridge over the waterway.

There are only three locks on the *Exeter Canal*. Apart from the sea lock
(Turf Lock) into the estuary and a disused side lock into the river at
Topsham, there is just the one lock, confusingly called Double Locks.
(The name could pinpoint the transition from the idea of two 'flash' locks
close together to the modern conception of a single 'pound' lock, with its
two sets of gates.) This lock has long, circular wooden balance beams.
Their great weight would doubtless have been necessary to offset the great
width of each lock gate. There is an inn beside the lock—The Double
Locks Hotel—which is almost, but not quite, as isolated as the Turf Hotel
at the canal's southern terminus.

SEEING THE EXETER CANAL

Apart from the historical interest pertaining to this ancient yet well‑main‑
tained waterway, the *Exeter Canal* is well worth visiting in conjunction
with the very fine Maritime Museum at the head of the canal. The Museum
features not only some remarkable and exotic craft afloat in the basin—
ranging from Arab dhows to a 130 year old steam dredger from the old
Bridgwater Dock—but also, inside, some smaller and even stranger craft—
including an old Bude Canal tub‑boat, a rectangular flat‑bottomed craft
incorporating four iron wheels for use on the inclined planes along that
canal. The museum occupies not only the warehouse at the head of the canal

but also the one just across the river, near the old Customs House—itself of passing interest.

Indeed, the best way to see the canal is probably to start at the museum at the Canal Head, then walk along the towpath (there is one each side all the way) to Topsham—about four miles. A ferryman in a rowing boat—which is the responsibility, as the *Exeter Canal* itself has always been, of the City of Exeter—takes people across the tidal Exe to Topsham. From here trains and buses provide a return journey to Exeter.

BOATING ON THE CANAL

The canal is of course accessible only from the sea. Although it is in very good condition there is now little traffic on it, and boats intending to enter from the Exe estuary at Turf Lock should give at least twenty-four hours notice of their intention to the canal authorities in Exeter (Exeter 74306). With its large locks and draught of over twelve feet (fourteen in places), the canal is well suited for yachts, and many use the canal head for laying up in winter. Rowing skiffs can also be hired from here in summer.

Bridgwater and Taunton Canal

Today a mere backwater, isolated from other canals and virtually unused, the *Bridgwater and Taunton* was once a canal with a grand future. For it was scheduled to be an integral part of a ship canal linking the Bristol and English Channels—an idea that fostered endless canal schemes in the South-West of England and which was never realized. The *Bridgwater and Taunton* was originally envisaged as part of such a line; but although the plan for the rest of the route 'stalled' at one stage or another, the *Bridgwater and Taunton* was completed to perform a useful function in its own right. The canal was opened in 1827, leaving the River Tone at Huntworth just south of Bridgwater and rejoining it at Taunton. Later (1841), the canal was extended from Huntworth to a large new dock on the river Parrett the other side of Bridgwater. But the canal was rather out on a limb, and was at least partly dependent for traffic on the short-lived *Grand Western* and *Chard* canals that were connected to its southern end and, like them, it succumbed in 1866 to the Bristol and Exeter Railway (later part of the Great Western). The canal traffic dropped from then on.

The canal has carried virtually no traffic since the beginning of this

234

Bridgwater Dock, Somerset.

century, although it received some attention from the army in the last war, when the swing bridges were fixed. Its locks have become unusable, Bridgwater Dock has been closed, and by rights the canal should have become a decaying ditch. But it hasn't—it is entirely 'in water' and full of fish, and although its lock gates are falling apart and the low bridges are fixed, it cannot be described as a derelict waterway. It is, in brief, a pleasant but isolated canal, more suited for fishing or walking than for boating.

The canal runs from Taunton to Bridgwater—a total of just over fourteen miles, in a generally north-easterly direction. It leaves the River Tone in Taunton at Firepool lock, and follows the course of the Tone as far as Creech St Michael, where the valley broadens out into the great flat plain of Sedgemoor. Closely accompanied by the main-line railway—the old Bristol and Exeter—the canal flows along the gentle slopes bordering the west side of the plain. There is a lock here and there and one or two small villages, like Durston and North Newton. It is a wide open landscape full of rich meadows, dykes and grazing cattle. The new M5 motorway makes its inevitable impact on the otherwise quiet agricultural scene, while the railway is never far off. Eventually, the canal enters Bridgwater, the 'new' loop curving in a slight cutting round the town before ending in a large dock which is now disused. At the far end of the dock is the River Parrett, a narrow and twisting tidal river.

PLACES TO SEE THE CANAL

It is unfortunate that the *Grand Western Canal* and the *Chard Canal,* which connected with the *Bridgwater and Taunton* at Taunton and Creech St Michael respectively, have been abandoned for so long. With their lifts, tunnels and inclined planes, they would both have been intriguing examples of canal engineering and operation. Today, they are of abiding interest to the hardened explorer of forgotten canals. There is little left to see of their junction with the *Bridgwater and Taunton.*

The latter is of more general interest. There are not many locks—a typical one is King's Lock at North Newton, which is now between the M5 and the railway line, but comfortably sheltered from each. As befits the high hopes that the canal originally reflected, the lock is broad like all the others— it also has concrete balance beams, decaying bottom gates, and interesting paddle gearing incorporating iron pulley wheels on top. The Taunton end is not very interesting as it has suffered from the encroachment of railway lines as well as the normal expansion of the town. Otherwise the canal's chief feature of interest is Bridgwater Dock. This was closed a few years ago and fixed concrete walls have replaced the tidal lock gates; but all sorts of relics still stand as witness to the dock's busy past. An old warehouse, a disused crane, rusting capstans, rotting lock gates and piles of river silt are all today's sorry evidence of a once prosperous little port.

Grand Western Canal

The title of this canal derives from its optimistic origins as part of a projected waterway to connect the English and Bristol Channels—a scheme which obsessed men of the West Country for decades and which never came to fruition. The *Grand Western* was in fact designed to extend the proposed Bristol–Taunton ship canal (which eventually came into being as the much more modest *Bridgwater and Taunton Canal*) from Taunton down to the Exe estuary. Failure of the overall plan scuppered the chance of an integrated waterway system in the south west and greatly reduced the chances of survival of the canals that would have been linked with it.

The *Grand Western Canal,* first conceived in 1792, suffered as much as any of the other canals from the uncertainties that surrounded the ship canal scheme. Only a portion of the main line was ever built, from Taunton to Loudwells, south-west of Wellington, and a ten mile branch from here to Tiverton. The main line, which was built on a dramatically different scale to its original plan and was in any case not opened until 1838 (twenty-four years after the Tiverton branch) was, like many other West Country canals, interesting for the ways in which it scaled the hilly countryside—in this case with a string of seven complicated vertical lifts and one unreliable inclined plane. Their sites are all of great interest to today's industrial archaeologists, but they made the canal cumbersome in operation and expensive to maintain. Thus, when the canal company was forced to give in to the Bristol and Exeter Railway Company that had been opened up not long after the canal's main line was completed, it was the Taunton–Loudwells section that the railway company abandoned as soon as possible, in 1867. Since then, the Tiverton branch has been the only surviving part of the *Grand Western Canal.*

A tiny local traffic in stone continued to use this branch for many years, but by the time the British Waterways Board inherited the waterway from British Railways in 1963, the branch itself had been abandoned. However, the canal has recently been restored to some kind of navigability; Devon County Council bought it from the BWB, invested time and effort therein,

Halberton Aqueduct. The *Grand Western Canal* has contrived to outlive this former branch of the Great Western Railway.

and now the waterway is designated a country park and lives again. There is even a horse-drawn pleasure boat running day trips from Tiverton Basin.

The *Grand Western Canal* as it exists today thus runs from Loudwells to Tiverton. It is in every way a pleasant, rural canal, situated in a lush green countryside well-loved by holiday-makers. This characteristic, and the fact that there are no locks along its length, doubtless played an important part in the decision to restore the canal.

The principal problem in restoration was sealing the leaks in the canal bed that had left the section around Halberton dry for many years, and the leaky aqueduct over the old railway south of Halberton. But a combination of old and new techniques—clay puddle and plastic sheeting—seems to have more or less cured these difficulties.

The canal's north-eastern terminus near Loudwells is in, as it were, the middle of nowhere, but near (coincidentally) the upper reaches of the River Tone, tucked away between the tight folds of hills that characterize the area. A cutting carries it southwards to the main Taunton–Exeter railway line and then sharply westwards, passing through a countryside that gradually opens out. The canal passes straight through the village of Sampford Peverell, whose church tower stands like a sentinel over the waterway. It passes the little wharf of Rock House and then circles round Halberton before crossing the disused Tiverton branch railway and entering Tiverton itself in a basin on the hillside high above the town. Just down the hill is a memento of the canal's undoing: a preserved Great Western Railway tank engine stands silently on a few yards of rails. It is ironic that while all the railways of Tiverton have now been dismantled, the much older canal should now live again.

PLACES TO SEE THE CANAL

All of the towpath of the *Grand Western Canal* is a public footpath, and it makes an excellent country walk, especially as the canal is so well stocked with all kinds of wildlife. There are no locks on the canal, but the tunnel at Loudwells is worth exploring—although it looks from the road above like no more than an underbridge. The canal fits well into the village of Sampford Peverell with its central wharf, and the Halberton Aqueduct is easily found beside a minor road. The aqueduct was built slightly skewed, its two iron arches encased in masonry and brick: the double width was employed in case the Bristol and Exeter Railway ever wished to double the track on its Tiverton branch. The aqueduct has long been a bugbear of the canal's team of maintenance men: it has been repaired recently, but it will doubtless continue to drip for ever. The horse-drawn 'Tivertonian' (as good a way as any of viewing the west end of the canal) operates from Tiverton Basin to a point near here.

BOATING ON THE CANAL

Short and isolated from other navigations, the *Grand Western Union* is obviously not cruising water, nor indeed suitable for any powered craft; the weeds would make short work of them. However, its clean, lockfree course is ideal for leisurely rowing or canoeing. Vegetation, both in and beside the waterway, thrives of course; but these are not just common weeds—in summer the surface of the water is richly clothed in glorious waterlilies.

13
Canals in Scotland

FORTH AND CLYDE CANAL

EDINBURGH AND GLASGOW UNION CANAL

CALEDONIAN CANAL

CRINAN CANAL

Scotland contains an interesting 'mixed bag' of canals, the most famous being the great *Caledonian Canal,* which links the seas from Fort William to Inverness and is still of great use to fishermen and yachtsmen—as is the much shorter *Crinan Canal,* on the West Coast. But there are others.

Within the Lowlands, there used to be quite a substantial network of interlinking waterways around Glasgow and Clydeside. These canals were built at the same period as most of the English canals, and often for similar reasons and on a comparable scale. Such were the *Monkland Canal,* built eastwards from Glasgow to tap mineral deposits at Monkland, south of Airdrie; and the *Glasgow, Paisley and Ardrossan Canal,* an unsuccessful venture to cut through from the Clyde at Glasgow to the sea at Ardrossan. (Its aims, and its failure, remind one of the *Great Western Canal* in south-west England.) Both these Scottish canals are now past praying for: the *G P & A*, long closed, was located within an entirely urban area and has, not surprisingly, been almost completely obliterated. The same goes for the *Monkland Canal,* which was closed in 1950. Most of its reaches within Glasgow and Coatbridge have been systematically redeveloped—filled in—to put space occupied by the canal to 'better' use—not an easy solution, since the *Monkland Canal* has always acted as a water feeder from reservoirs east of Airdrie to the summit level of the *Forth and Clyde Canal.* The water has had to be piped along the bed of the *Monkland.* But piping or not, there is now little to see of the canal in the urban areas it passes through, although isolated stretches of the canal out of town can be explored.

240

Forth and Clyde Canal

The recipient of the *Monkland's* waters was and is the *Forth and Clyde Canal*. Although this is now closed to navigation, with most of its bridges culverted and its locks dismantled, the *Forth and Clyde* is still full of water—it supplies local industries with water for cooling purposes—and is largely intact as a continuous line. It is worth a brief appraisal here.

The *Forth and Clyde Canal* runs from Bowling, on the north bank of the Clyde, past Glasgow and through Kirkintilloch, Kilsyth and Falkirk to the mouth of the River Carron on the south side of the Forth estuary. The canal's dimensions are generous by English standards, but were regarded very much as a compromise when the canal was built. For at the time there had been much interest in the idea of a small ship canal across Scotland (more on the *Caledonian Canal* scale). But as built, the *Forth and Clyde* was all of seven feet deep, and this was deepened later. The locks were twenty feet wide; and the avoidance of fixed bridges gave virtually unlimited head-room—essential for the tall masts of the fishing boats for which the canal mainly catered.

The canal was built with nineteen locks up from Bowling Basin to a sixteen-mile long summit level. This follows the contours on the south side of the shallow valley of (mainly) the River Kelvin. East of Kilsyth is the first of four locks down; then at Falkirk is a steep flight of locks right down to docks at Grangemouth. Apart from linking the Forth to the Clyde, the canal had a spur two and three quarter miles long into Glasgow, to meet the *Monkland Canal;* and it was formerly joined in Falkirk by the *Edinburgh and Glasgow Union Canal*—a junction which has long since disappeared.

The builders of the *Forth and Clyde* did not do things by halves. The canal sweeps along through the undulating landscape quite unhindered by topographical irregularities. Hence the big embankments and cuttings, and the remarkable number of aqueducts large and small by which it steps over

The junction of the *Forth and Clyde* with the River Clyde.

Kelvin Aqueduct, Glasgow, on the *Forth and Clyde Canal*.

the innumerable roads, railways and rivers that intersect its course. Great solid masonry structures, these aqueducts are found along the canal in numbers quite unheard of on English canals. They are all the more impressive for carrying a canal which, although closed as a through route, is still very much full of water.

THE CANAL TODAY

The *Forth and Clyde* has been officially closed to navigation since 1962. It was undoubtedly the swing bridges that were the cause of the canal's undoing, for the bottlenecks and delays caused to motor traffic along a line right across Scotland's crowded Lowlands, led inevitably to the replacement of the bridges by low fixed structures. Once started, this process was irreversible and the demise of the canal was assured. Now, the only boats that can use the canal's deep waters are portable craft that can be carried round the modern obstructions. It is also very suitable for walkers, and very accessible by car. Principal points of interest along the canal, from west to east, are the terminal basin at Bowling, which is still very much in use as a sheltered mooring for yachts just off the Clyde; the Kelvin Aqueduct in Glasgow, at the foot of the locks in Maryhill; the aqueduct over the Luggie Water in Kirkintilloch; and the long flight of locks in Falkirk, with canal stabling and warehousing. All these are now the principal but useless features in a landscaped urban parkland. There also remain the extensive docks that mark the canal's eastern terminus at Grangemouth.

Edinburgh and Glasgow Union Canal

The 'Union' Canal was built to connect the *Forth and Clyde Canal* to the city of Edinburgh. Like the latter, it is now still very much 'in water' over

Almond Aqueduct, on the *Edinburgh and Glasgow Union Canal,* is one of several massive aqueducts in the Lowlands.

virtually the whole of its length, but it is likewise closed to navigation and suffering from ever-increasing numbers of its bridges being culverted.

The canal is all on one level for thirty-one miles from Glasgow to just above Falkirk. There used to be a steep flight of broad locks here to join the *Forth and Clyde,* but these have completely disappeared and the canal suddenly stops dead on the hillside. The *Union Canal* was built as a barge canal so, unlike the *Forth and Clyde,* it is crossed by fixed bridges. (These

243

are classic stone arched bridges somewhat reminiscent of those on the *Lancaster Canal*.) There is also a tunnel under the hill at Falkirk, a spacious affair complete with a towpath.

But the best feature of the *Union Canal* is the aqueducts. There are fewer than on the *Forth and Clyde*, but its biggest ones are gigantic and make most English aqueducts look very small indeed. They are ample testimony to the canal's late construction, and make it all the more regrettable that it should ever have been abandoned. The biggest is the Avon Aqueduct, a twelve-arched masonry structure over 800 feet long, carrying the canal eighty feet over the valley floor. The other two principal ones, over the Waters of Leith near Edinburgh, and over the River Almond a few miles to the west, are much shorter but almost as high and very impressive. The Almond Aqueduct has a beautiful setting in a secluded river valley. A feeder stream continues to flow into the canal at one end of the aqueduct.

Caledonian Canal

The best known of Scotland's canals is of course the *Caledonian*. This is Scotland's biggest canal, in terms of both its length and the size of the vessels it will carry. In fact, the *Caledonian Canal* is not simply a canal but a series of canal sections linking up various deep inland lochs to form a sea-to-sea waterway. It is none the less impressive for that. The dramatic staircase flights of locks—all of which are now fully mechanized—the very scale of

The eastern end of the *Caledonian Canal* reaches well out into the Beauly Firth, to ensure sufficient draught for vessels.

Laggan Locks on the mighty *Caledonian Canal*.

the canal, its continued commercial use, and the sheer grandeur of the scenery through which the navigation passes, all combine to give the *Caledonian* the stamp of a very special waterway.

The canal is unusual also in having been promoted and financed by the Government. The project was designed principally to provide a short and safe alternative route for shipping that would normally have to make the difficult passage right round the north of Scotland. The military value of this at a time of war against Napoleon was an important factor in the decision to build the canal along a route that was obviously attractive in containing a string of long but disconnected lochs that make up the dead straight line of Glen More. The canal was engineered by Thomas Telford. Its construction, in a wild and remote part of Britain far from sources of materials and expertise, was a long drawn-out affair; and the commercial prospects for the canal declined with the proliferation of steam-powered boats that were, inevitably, less reluctant to face Cape Wrath's notorious winds and currents than were sailing boats.

The canal stretches from Corpach, by Fort William, to Clachnaharry, near Inverness. It is a very grand canal framed by grand scenery, with steep and barren mountains rising on either side of the long straight glen where the virtually impenetrable land mass encloses the soft green scenery of the valley floor. The names of the towns along the way—Fort Augustus and Fort William—serve as a reminder of the military importance of the glen as a principal line of defence in the troubled times following the Jacobite rebellions. Along the glen there are three fresh water lochs—Lochy, Oich and Ness. These are joined to each other and to the tidal waters at each end by four separate canal cuts. It is these cuts that form the actual canal.

Now hydraulically operated, the lock gates at Fort Augustus used to be opened by turning the capstan in the foreground.

THE COURSE OF THE CANAL

There are twenty-nine locks on the canal, including the sea locks at each end—the one at Clachnaharry had to be built well out into the Beauly Firth to reach deep enough water. The lochs provide forty miles of navigable waterway (Loch Ness alone is twenty-four miles long), while the canal cuts account for the remaining twenty miles. There are other isolated locks in places along the waterway, but most of them are grouped in steep staircase flights—the two at Corpach quickly followed by eight at Banavie (Neptune's Staircase), two at Laggan, five at Fort Augustus and four at Muirtown, at the eastern end of the canal. These locks, over 150 feet long and thirty-five feet wide, hardly place the *Caledonian Canal* in the ship canal category today, but their size must have seemed startling to the Highlanders when the canal was opened in 1822, and the great staircases are still—rightly—a source of wonder for the many tourists who visit the canal today. (Only the mainte-nance men are able to appreciate the actual depth of these great locks—which have to offer boats at least fourteen feet over the bottom sill.) The gates and sluices are operated entirely hydraulically, although many of the old hand-operated capstans can still be seen beside the locks. These are smartly painted in black and white—well maintained like most of the canal structures—although the locks lack the bold and useful numbers that characterize the *Crinan Canal* locks. There are of course no balance beams on the *Caledonian* locks.

It is hardly surprising that there is only one aqueduct on the canal—to take it over the River Lochy. But another greater work on the canal is the summit level above Laggan Locks. To make any form of cutting for a canal as wide and deep as this one must have been a laborious job. The summit level itself stretches from Laggan Locks to Culloch Lock, embrac-ing Loch Oich, which is supplied mainly by the River Garry at Invergarry. Depth of water within the loch varies wildly, and the channel through it is carefully buoyed.

The bridges on the canal are all swing bridges—most of them double-tracked structures of sufficiently comfortable proportions not to delay or dismay motorists, except of course, when they open to shipping. There are no fixed bridges at all across the canal, so that headroom is limited only by the very tall high-tension cables that cross the canal in several places.

Muirtown Locks, *Caledonian Canal,* at dusk. The white paint helps ships' pilots to avoid damaging the gates.

The *Caledonian Canal* is very well endowed with scenic attractions before one even turns to the canal works themselves. With Ben Nevis towering 4400 feet above Fort William, and with the well-publicized mystery of Loch Ness towards the other end (as well as the glorious views along the deep glens) the area is already well-exploited for its tourist potential. The canal itself is inevitably—and deservedly—drawn into the tourist's net.

Largely because of what is still called 'General Wade's Military Road' along the Great Glen, the canal is very accessible to road traffic to most points of interest. Chief among these are the flights of locks, where to watch a sizeable fishing vessel or small coaster working its way up or down the staircase is always a fascinating sight. The eight Banavie Locks are just by Corpach, and there are road and rail swing bridges at the foot of the locks, just for good measure. The last three locks on the canal, with the small entrance basin and the tidal waters of Loch Linnhe, are only a mile away. Fort Augustus Locks, lowering the canal into the waters of Loch Ness, were once flanked by rail and road bridges; but now the old railway swing bridge at the head of the locks has gone. The other main focus of interest on the canal is at Muirtown, Inverness, where a large basin at the foot of the flight is often busy with small coasting vessels; and the canal's promontory at nearby Clachnaharry out into the deep waters of the Beauly Firth is also worth a look. The old headquarters of the Caledonian Canal Commission on the main road have a marble inscribed monument to the Canal's architect, Thomas Telford, composed by the poet Robert Southey.

BOATING ON THE CALEDONIAN CANAL

The *Caledonian Canal* is not only a first-class sea-to-sea route but also a waterway worth dawdling over. There are plenty of anchorages in the lochs and plenty of places to visit—and all framed by superb mountain scenery. In recognition of this, several hire firms have now begun to operate on the waterway.

Negotiating the waterway, especially in fine weather, is an exhilarating experience, with a constant switching from the confines of a canal cut to the great open stretches of the lochs. The canal itself is plain sailing: all the bridges and locks are operated by keepers. The locks are all electrified now, so progress is rapid even through the staircase flights. The lochs are wide and —except for Loch Oich—extremely deep; but the sixty-mile gulley that contains the route can sometimes act as a funnel for a strong wind straight up or down the lochs. This can get up and die down at very short notice.

Dunardry Locks on the *Crinan Canal*, which climbs sharply up the edge of the Knapdale Forest to its short summit level. Almost all of these locks are still manually operated.

Crinan Canal

The *Crinan Canal* is a nine mile canal forming a short cut across the head of the Kintyre peninsula. Built primarily for the benefit of fishing boats, the canal provides a ready communication between the Sound of Jura—and thus the Atlantic—on one side, and Loch Fyne—and hence the Firth of Clyde—on the other; over eighty miles round the Mull of Kintyre are saved by this route. It must have been somewhat frustrating to the canal's builders to see how much easier it would have been to cut through the isthmus further south, at Tarbert.

The canal was opened in 1801, and although it was of little success in commercial terms it was indeed a boon to the fishing industry. And today it remains a great asset not only to fishermen but to ever-increasing numbers of yachtsmen keen to save time and fuel, or to take up a sheltered mooring on a rare and delightful little waterway.

The canal traverses landscape where one would hardly expect to find inland waterways. Mountains and mud flats at the western end, with steep slopes clad in conifers rising on either side of the canal's summit level, and then the journey along and above the shores of Loch Fyne, are hardly typical backdrops for a navigable canal. The canal has a good towpath but is anyway accompanied throughout nearly all of its length by public roads, so it is easy to appreciate by car. Starting at Ardrishaig, the canal's eastern end is marked by the first of seven steel swing bridges. This one has the distinction of being driven by compressed air, while the lock beneath it is also power-operated. Apart from this and the two locks at Crinan itself, all the locks are still hand-operated. The long wooden balance beams, neatly painted in black and white and boldly numbered in big iron figures, stand out in their Highland surroundings.

There are four locks at Ardrishaig and a second bridge, then a three mile level pound along the contours to Cairnbaan. Here the forested hills close in, and the canal climbs a short flight of four locks to the summit—a startlingly short pound only just over half a mile long. (The water level is maintained by eight small supply lochs hidden away in the craggy hills to the south.) Then almost immediately the first of the five Dunardry Locks is encountered, dropping the canal down to another long level. A stretch of low-lying moorland traversed by the tidal River Add opens up to the north, while almost vertical mountains contain the canal's southern flank. At the little village of Bellanoch the normally narrow waterway broadens out briefly, providing an excellent mooring place for boats, but from here to Crinan the canal hugs the cliffs overlooking the sea and becomes extremely narrow—too narrow for two boats to pass. Finally it arrives at the tiny haven of Crinan, where a small basin between the two lochs provides snug moorings for fishing boats, motorboats and yachts alike.

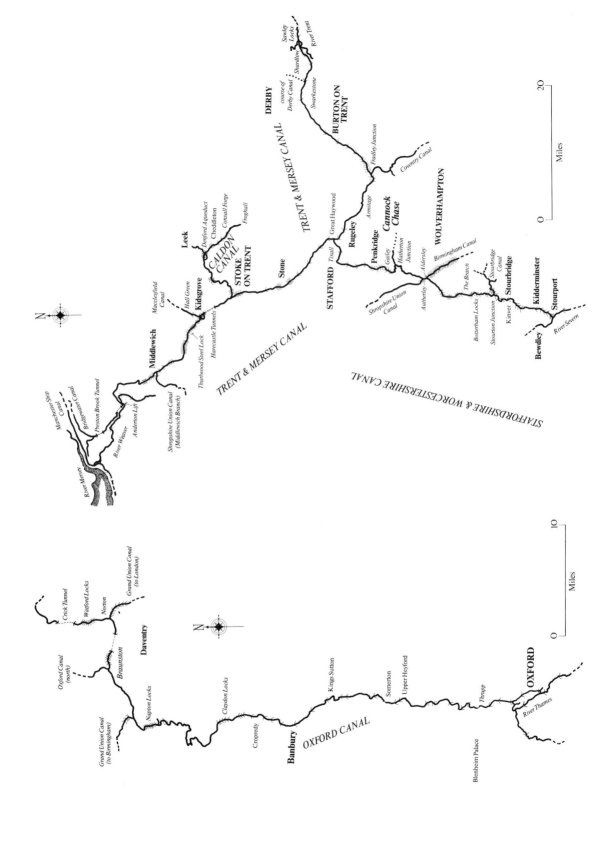

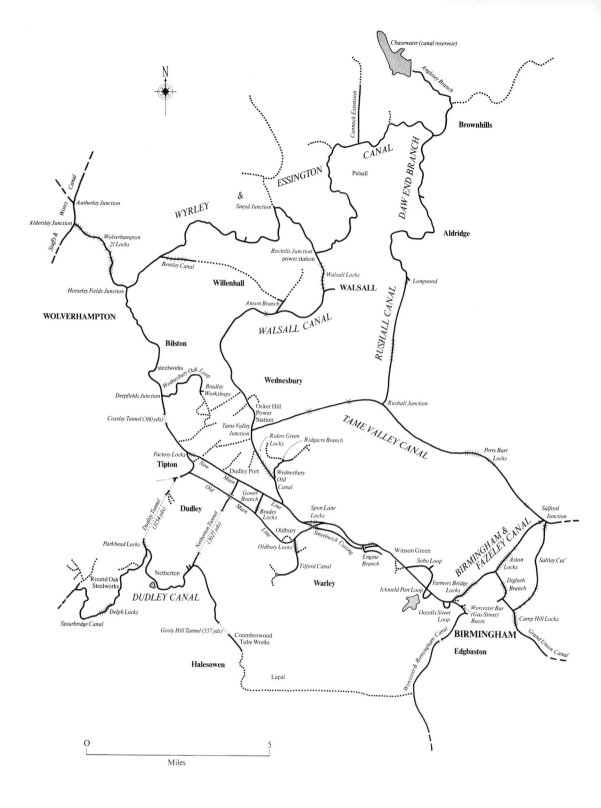

N

Chasewater (canal reservoir)

Angleey Branch

Brownhills

Cannock Extension

ESSINGTON CANAL

DAW END BRANCH

Pelsall

WYRLEY

&

Sneyd Junction

Staffs & Worcs Canal

Autherley Junction

Aldersley Junction

Wolverhampton 21 Locks

Aldridge

Bentley Canal

Birchills Junction power station

Walsall Locks

Longwood

Willenhall

WALSALL

Horseley Fields Junction

RUSHALL CANAL

WOLVERHAMPTON

Anson Branch

WALSALL CANAL

Bilston

steelworks

Wednesbury Oak Loop

Wednesbury

Deepfields Junction

Bradley Workshops

Coseley Tunnel (360 yds)

Ocker Hill Power Station

Rushall Junction

Factory Locks

Tipton

New

Tame Valley Junction

Riders Green Locks

Ridgacre Branch

TAME VALLEY CANAL

Perry Barr Locks

Dudley Port

Old

Main

Wednesbury Old Canal

Dudley Tunnel (3154 yds)

Dudley

Gower Branch

Main

Brades Locks

Spon Lane Locks

Salford Junction

Netherton Tunnel (3027 yds)

Line

Oldbury

Smethwick Cutting

Winson Green

Aston Locks

Saltley Cut'

Parkhead Locks

Netherton

Line

Oldbury Locks

Engine Branch

Soho Loop

BIRMINGHAM & FAZELEY CANAL

Digbeth Branch

Round Oak Steelworks

Titford Canal

Farmers Bridge Locks

Icknield Port Loop

DUDLEY CANAL

Warley

Oozells Street Loop

Worcester Bar (Gas Street) Basin

Camp Hill Locks

Delph Locks

Stourbridge Canal

Gosty Hill Tunnel (557 yds)

Coombeswood Tube Works

Worcester & Birmingham Canal

BIRMINGHAM

Grand Union Canal

Edgbaston

Halesowen

Lapal

O   5

Miles

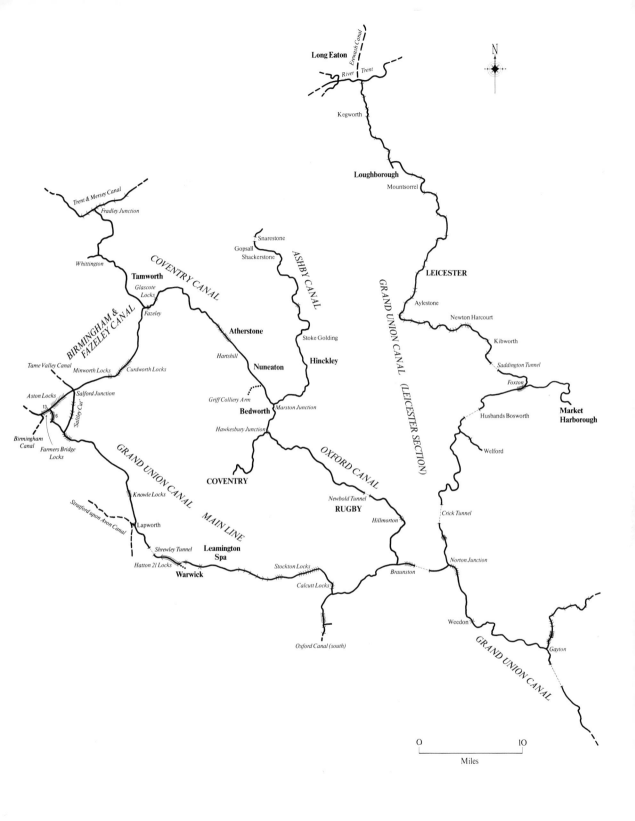

N

Long Eaton

Erewash Canal

River *Trent*

Kegworth

Loughborough

Mountsorrel

Trent & Mersey Canal

Fradley Junction

Whittington

COVENTRY CANAL

Tamworth

Glascote Locks

Fazeley

Snarestone

Gopsall
Shackerstone

ASHBY CANAL

LEICESTER

Aylestone

Newton Harcourt

Kibworth

Saddington Tunnel

GRAND UNION CANAL (LEICESTER SECTION)

Foxton

BIRMINGHAM & FAZELEY CANAL

Atherstone

Stoke Golding

Hartshill

Nuneaton

Hinckley

Tame Valley Canal

Minworth Locks

Curdworth Locks

Aston Locks

Salford Junction

13

6

Saltley Cut

Birmingham Canal

Farmers Bridge Locks

Griff Colliery Arm

Bedworth

Marston Junction

Hawkesbury Junction

COVENTRY

Market Harborough

Husbands Bosworth

Welford

OXFORD CANAL

GRAND UNION CANAL MAIN LINE

Knowle Locks

Newbold Tunnel

RUGBY

Crick Tunnel

Hillmorton

Stratford upon Avon Canal

Lapworth

Shrewley Tunnel

Leamington Spa

Norton Junction

Hatton 21 Locks

Warwick

Stockton Locks

Calcutt Locks

Braunston

Weedon

GRAND UNION CANAL

Gayton

Oxford Canal (south)

O   IO

Miles

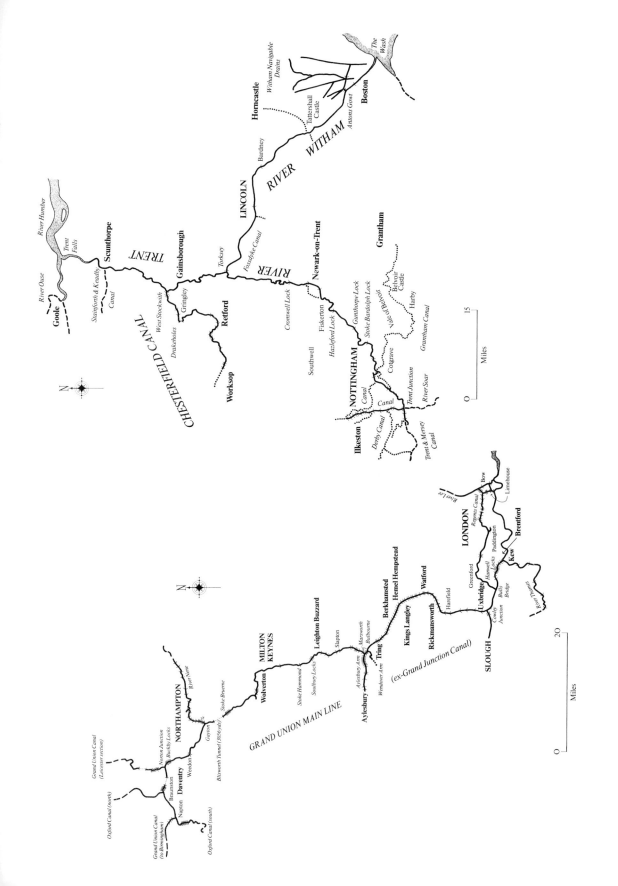

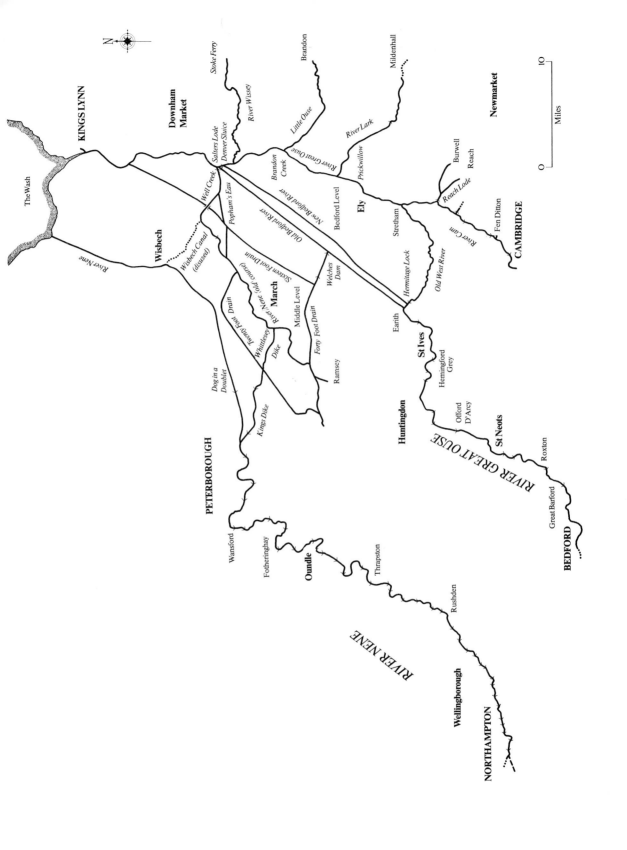

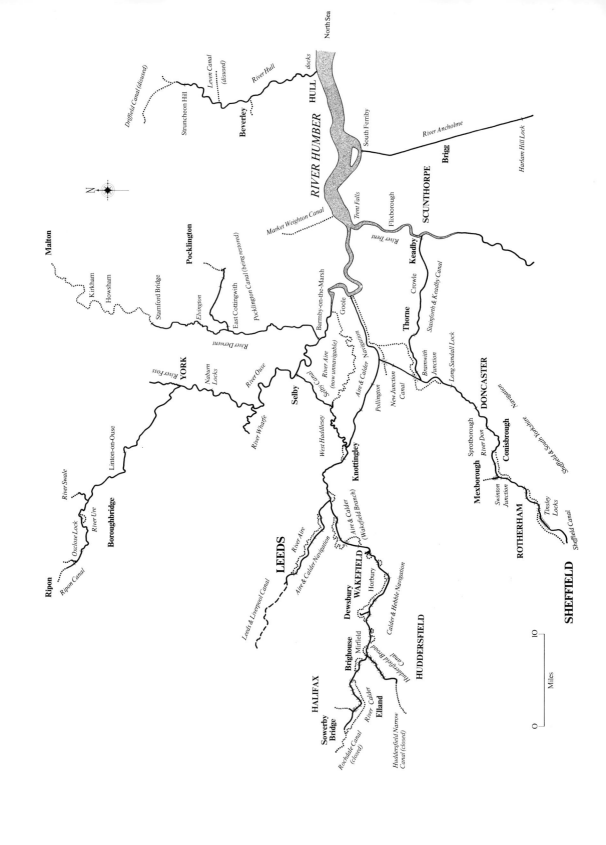

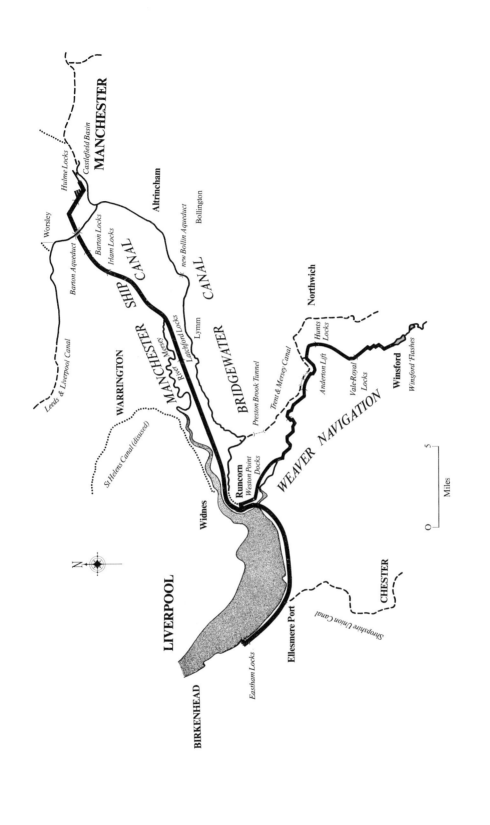

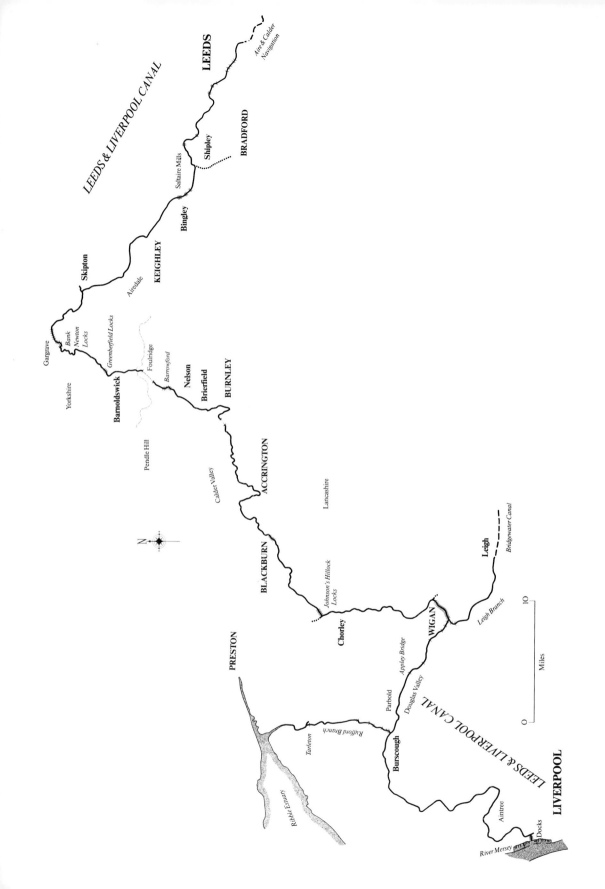

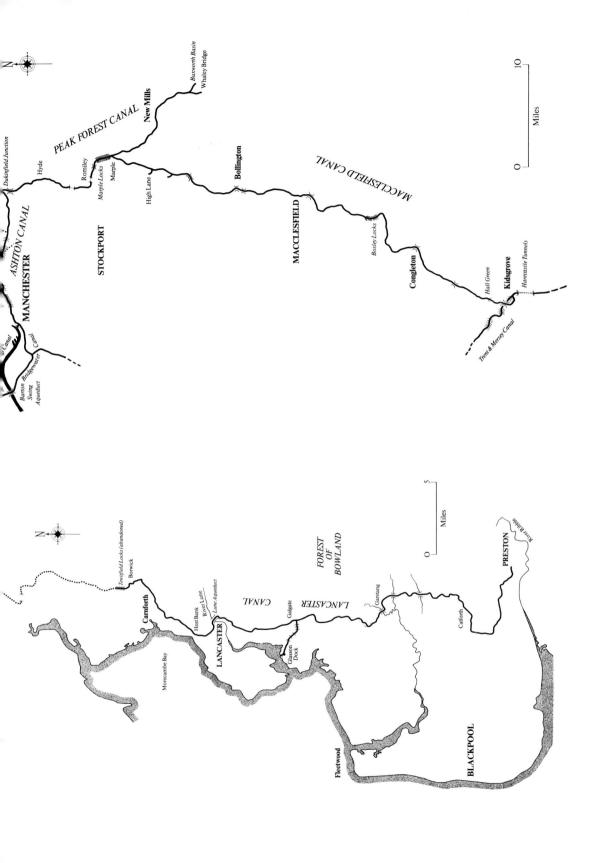

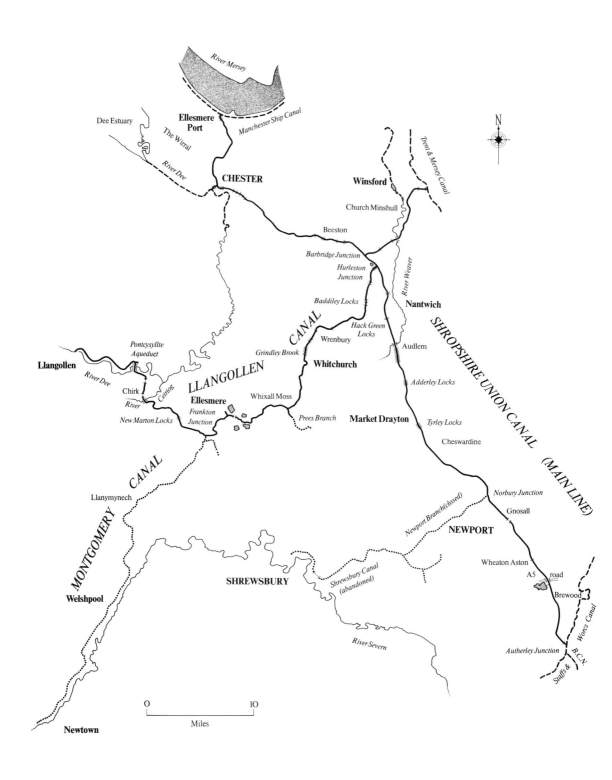

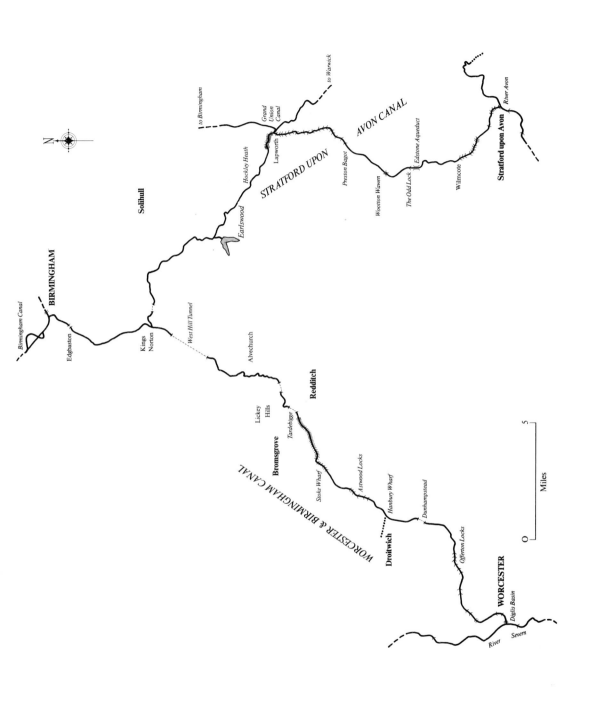

N

BIRMINGHAM
Birmingham Canal
Edgbaston

Solihull

Hockley Heath
Earlswood
Kings Norton
West Hill Tunnel

STRATFORD UPON
Lapworth
Grand Union Canal
to Birmingham
to Warwick

AVON CANAL
Preston Bagot
Wootton Wawen
The Old Lock
Edstone Aqueduct
Wilmcote
Stratford upon Avon
River Avon

Alvechurch
Lickey Hills
Redditch
Tardebigge
Bromsgrove
Stoke Wharf
Astwood Locks
Hanbury Wharf
Dunhampstead

WORCESTER & BIRMINGHAM CANAL
Droitwich
Offerton Locks

WORCESTER
Diglis Basin
River Severn

Miles
0   5

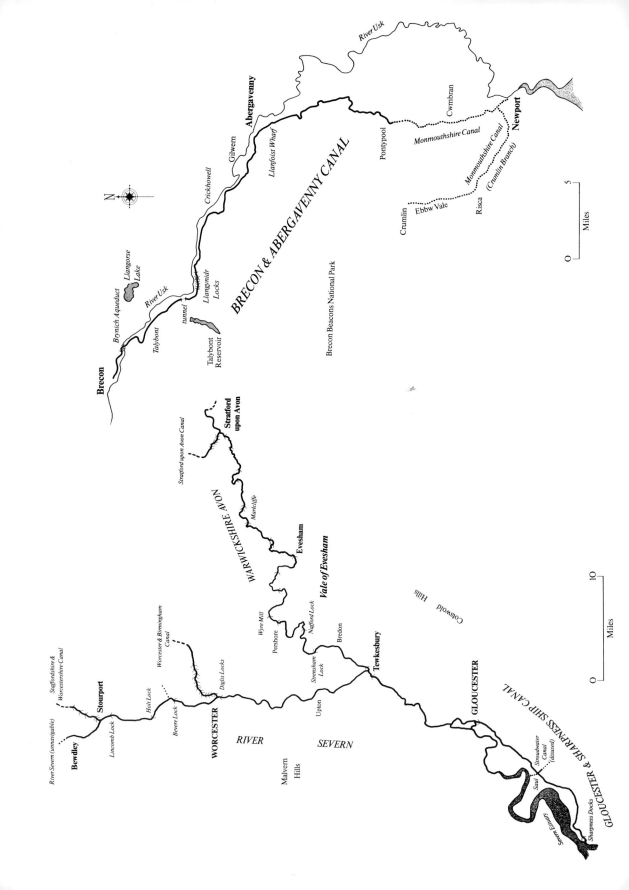

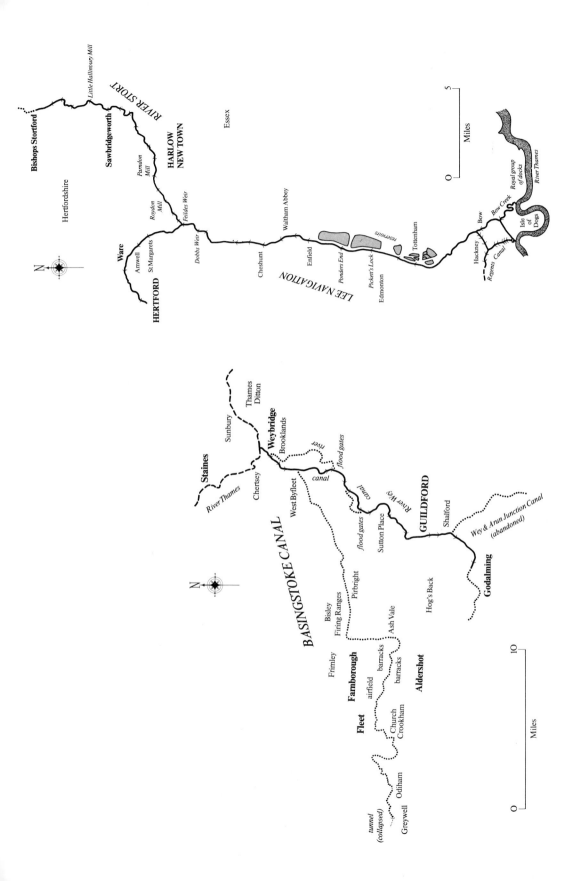

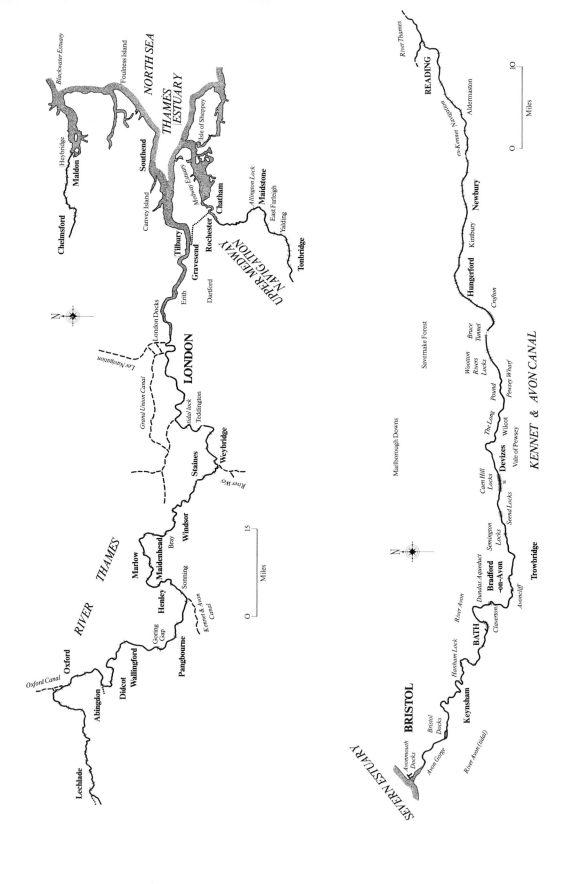

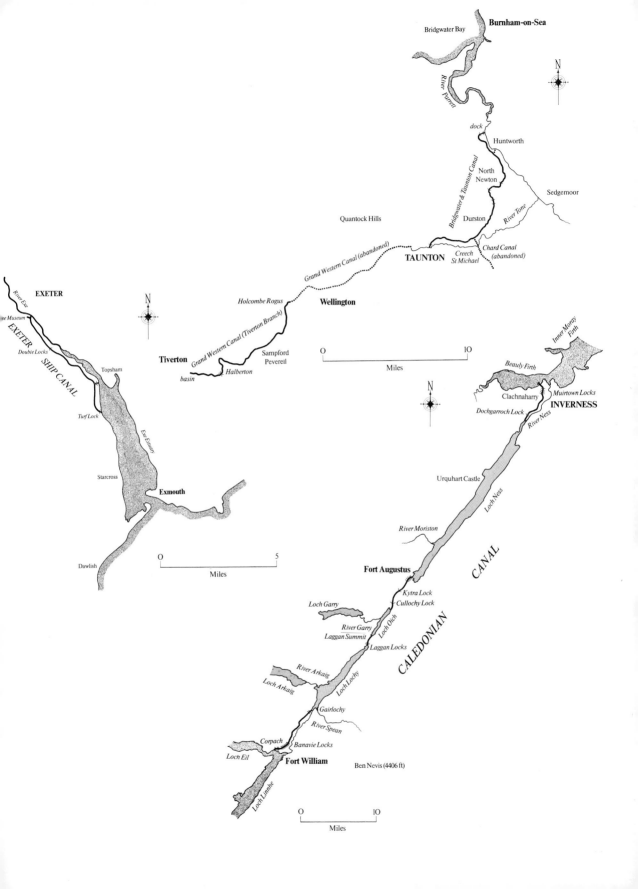

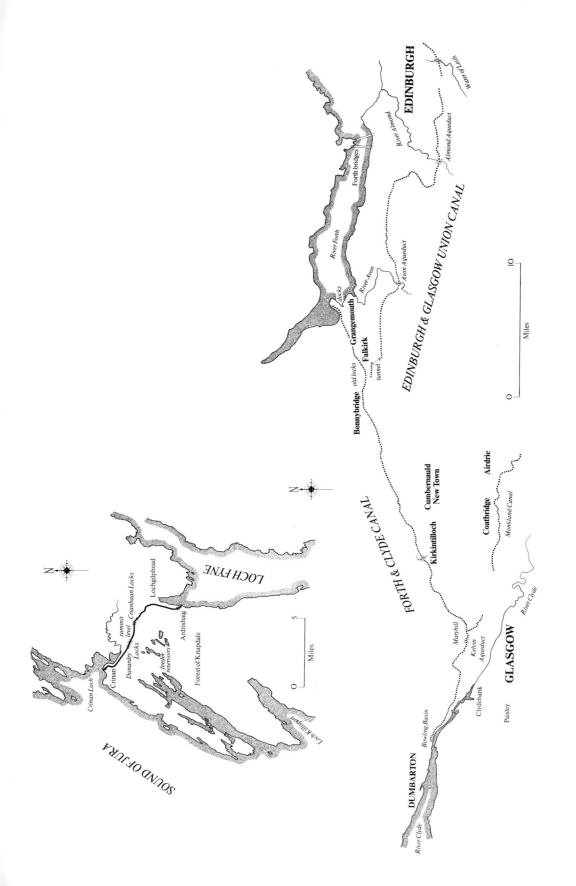

SOUND OF JURA

Crinan Loch

Crinan

summit
level
Dunardry
Locks
Crianban Locks

feeder
reservoirs

Forest of Knapdale

Lochgilphead

Ardrishaig

LOCH FYNE

Loch Killisport

N

0 Miles 5

EDINBURGH

Water of Leith

River Almond

Almond Aqueduct

Forth bridges

River Forth

docks

Grangemouth

River Avon

Avon Aqueduct

Falkirk

tunnel

Bonnybridge

old locks

EDINBURGH & GLASGOW UNION CANAL

FORTH & CLYDE CANAL

Cumbernauld
New Town

Airdrie

Coatbridge

Monkland Canal

Kirkintilloch

Maryhill

Kelvin
Aqueduct

GLASGOW

River Clyde

Bowling Basin

Clydebank

Paisley

DUMBARTON

River Clyde

N

0 Miles 10

Index